e-Services

Edited by Sam Goundar

Published in London, United Kingdom

IntechOpen

Supporting open minds since 2005

e-Services
http://dx.doi.org/10.5772/intechopen.73471
Edited by Sam Goundar

Contributors
John Sunday Sunday Ojo, Suppanunta Romprasert, Jirayut Monjagapate, Tetsuro Saisho, Samuel Owusu-Ansah, Stephen Budu, Rosina Akuokor Anati Budu, Keiko Tsujioka, Emmenual Vimlesh Reddy, Sam Goundar, Reginald Gani, Sharika Kumar, Fatemeh Saber, Samuel Kyobe, Marlene Amorim, Mario Rodrigues, Carla Gomes, Kunal Alvin Lal, Ashmita Chand, Pranav Vyas, Oliver Igwebuike Abbah, Uzoamaka Ogwo

First published in London, United Kingdom, 2021 by IntechOpen
IntechOpen is the global imprint of INTECHOPEN LIMITED, registered in England and Wales, registration number: 11086078, 5 Princes Gate Court, London, SW7 2QJ, United Kingdom
Printed in Croatia

British Library Cataloguing-in-Publication Data
A catalogue record for this book is available from the British Library

Additional hard and PDF copies can be obtained from orders@intechopen.com

e-Services
Edited by Sam Goundar
p. cm.
Print ISBN 978-1-78985-508-1
Online ISBN 978-1-83880-791-7
eBook (PDF) ISBN 978-1-83880-792-4

For EU product safety concerns:
IN TECH d.o.o., Prolaz Marije Krucifikse Kozulić 3, 51000 Rijeka, Croatia, info@intechopen.com or visit our website at intechopen.com.

Meet the editor

Professor Dr. Sam Goundar is an international academic having taught at twelve different universities in ten different countries. He is editor-in-chief of the *International Journal of Blockchains and Cryptocurrencies*, *International Journal of Fog Computing*, and *International Journal of Cloud Applications and Computing*. He is a section editor for the *Journal of Education and Information Technologies*. He is also on the editorial review boards of more than twenty high-impact factor journals. To date, Dr. Goundar has 101 publications to his credit, including journal papers and book chapters. He is currently working (writing/editing) on twelve book projects, eight of which have been published already and four that are expected to be published in 2021.

Contents

Preface

The turn of the new millennium has seen an explosion of activity around electronic services (e-Services) in the form of e-commerce, e-business, e-government, e-learning, and so on. The provision of all possible goods and services electronically via the Internet with the use of semantic web technologies has seen a paradigm shift from the traditional brick-and-mortar location-based services to the ubiquitous provision of goods and services. An understanding of this paradigm shift and the fundamental properties of e-service composition is required in order to take full advantage of the paradigm.

E-services are interactive web applications where the user's interaction, provision, and access to services is enabled via the Internet and the user's web browser. Many organizations and governments recognize e-services as assets of information, consisting of re-engineered business processes that use information communications technologies and web applications to generate new revenue streams, create efficiencies, and provide global services. This book is a reference book for scholars, researchers, and practitioners looking to update their knowledge on methodologies, theoretical analyses, modeling, simulation, and empirical studies on e-services.

Chapters on the evolving theory and practice related to e-services, including e-business, e-government, e-commerce, and e-learning provide comprehensive coverage and understanding of the use of e-services within the technological, business, management, and organizational domains. The book also provides a thorough examination of e-services with respect to issues of management, governance, trust, privacy, and interoperability. It explores various e-services related to commerce, business, health, learning, government, and educational technology. It provides insight into the relevance of educational technology, e-commerce solutions, and the use of e-services in the public and private sectors.

This edited book is divided into eleven chapters that present introductory concepts of e-services as well as examine various applications of e-services.

I am proud to present this book on e-services. I would like to thank all the chapter reviewers as well as the administrative and editorial support staff of IntechOpen, especially Mr. Josip Knapić, who have ably supported me in getting this book to press and publication. Finally, I would like to thank all the contributing chapter authors. Without their submissions, tireless efforts, and contributions, this book would not be possible.

I hope everyone will enjoy reading and learning from this book. I hope it will inspire and encourage readers to start their own research on e-services.

I congratulate everyone involved in the writing, reviewing, editing, and publishing of this book.

Please contact sam.goundar@gmail.com with any comments or questions.

Dr. Sam Goundar
Professor,
RMIT University,
Hanoi, Vietnam

Chapter 1

Introductory Chapter: Introduction to e-Services

Sam Goundar

1. Introduction

e-Services have been around for a while. For example, according to Lewis [1] "what likely counts as the very first legitimate online transaction goes to Dan Kohn in August 1994", who created a website called NetMarket, and on August 11, sold a CD of Sting's "Ten Summoner's Tales" to a friend in Philadelphia. Starting with this first e-Commerce transaction almost 30 years ago, we have come a long way. This includes the making of the richest man in the world today, Jeff Bezos and his pioneer e-Commerce startup, "Amazon". Coppola [2] from Statista. Com predicts that the global retail e-Commerce sales worldwide would reach 6.39 trillion USD and account 21.8% of share of total global retail sales by 2024. Security of Internet transactions was a major big issue at that time, but "the release of Netscape 1.0 in 1994 featured a protocol called Secure Socket Layer (SSL) that kept both the sending and receiving side of an online transaction secure. SSL made sure that personal information could be encrypted on the web. The first third-party credit card processing companies were launched shortly after. This made the first ever secure online retail transaction possible".

After e-Commerce, came other e-services such as e-Learning, e-Health, e-Banking, e-Agriculture, e-Government, etc., which all have now moved onto the mobile device's platforms (especially smartphones) and renamed to m-Commerce, m-Learning, m-Health, m-Banking, etc. Many of these e-Services survived the dotcom bubble or the dotcom boom, which saw the increase of such services and domain name registrations from 1996 and then crash from 2001. "During the crash, many online shopping companies, such as Pets.com, Webvan, and Boo.com, as well as several communication companies, such as Worldcom, NorthPoint Communications, and Global Crossing, failed and shut down". The value of e-Services was further realized once the governments got on the bandwagon and introduced e-Government and e-Governance to provide government services to its citizens as well as show transparency in governance. "As growth in the technology sector stabilized, companies consolidated; some, such as Amazon.com, eBay, and Google gained market share and came to dominate their respective fields". For example, in 2021, during Amazon Prime Day (for Amazon Prime Members), the one-day worldwide sales were 10.4 billion USD and Amazon sold 175 million items ([3], Statista.Com).

The usage and relevance of e-Services is more significant today than it ever was. With the use of technology for all services and the forced digital transformation of our work and lives because of the COVID-19 pandemic, the year 2020 and 2021 has seen more services being offered online/electronically (e-Services) than ever in its history. Because of social distancing requirements of the coronavirus, businesses closed doors and started transacting online, schools and universities started teaching online (e-Learning), food was being ordered and delivered via e-platforms, fitness trainers and gyms started having video training sessions with their clients, and we all adjusted our schedules to work from home and live-in isolation within our own or our family bubble. And the trend is going to continue in years to come. We are not going back to pre-2020 normal days. This is the new normal – working from home and using e-Services is the way things are going to be. And why not? Using e-Services saves costs and saves lives. We have realized that almost all the things that we were doing in the physical world can be done in the digital world. According to the World Trade Organization [4], COVID-19 has "resulted in spikes in business-to-consumers (B2C) sales and an increase in business-to-business (B2B) e-commerce.2 The sales is particularly evident in online sales of medical supplies, household essentials and food products".

e-Services have existed for decades and enabled organizations to gain competitive advantage while integrating their functions and streamlining their operations. Over the years, e-Services have evolved with developments in technology and new ways of doing business. Technologies like cloud computing, in particular, have made access to e-Services ubiquitous. Such technologies were needed to cater for new business models like e-Commerce, e-Learning, e-Government, etc. that resulted in a different business environment. This rapid acceleration of technological diffusion has changed the way we work, do business, and live. The business and research community took this interplay of work, technology and peoples into serious consideration which generated many research and publications on e-Services and socio-technical systems. The current e-Services that we have are results of these research and development.

Today' Fourth Industrial Revolution (4IR), is building on the Third (the digital revolution), and Future of Work is all about automation, machine learning and artificial intelligence. The First Industrial Revolution introduced mechanization, which saved us from mundane tasks such as fetching water from the well and collecting firewood. The Second Industrial Revolution used electrification to power factories into mass production and its products again assisted us for washing clothes, keeping our drinks cold and heating our food. The Third Industrial Revolution was about Information Communications Technology to automate office work and business processes. However, we were still required to start the computers, run the programs, check the results, monitor processes, and take corrective action. With the Fourth Industrial Revolution, the intention is to completely automate, let machines handle everything and totally free ourselves.

In terms of technology, we initially started off by doing our own developments and isolating ourselves from others because we were out to compete, take market share as the bottom line was vital. For example, we started off with different and incompatible computer hardware in the form of distinct models where work done on one computer could not be used on another. Then, we went through the platform wars of Microsoft vs. Apple vs. Linux, etc. And the same is being witnessed in the mobile platforms and Apple and Android. However, we have come to our senses and realized it is more profitable to become compatible, interoperable and converge. Technologies have converged, likewise. Every existing and new technology needs to be able to converge with others for survival. The Internet became the platform for technological convergence and a one-stop shop to find anything and everything.

Your smartphone is an ideal example of technical convergence. Initially, with a phone, you could only make voice phone calls. Now, with a smartphone, not only can you make a voice phone call, but you can also text, email, take photos, record videos, listen to music, make payments, get a ride, do banking, order food, browse the internet, and the list is endless. Just imagine how many different technologies have converged into your smartphone: phone (voice call), camera (take photos), walkman (listen to music), video recorder (record videos), computer (email, browse, etc.) and the list of technologies that have converged into your smartphone goes on. Similarly, a number of technologies have converged with e-Services. For example, blockchains and smart contracts can now be used to converge different supply chains of all the stakeholders of a particular product, ensure transparency and build trust with customers and each other.

Organizations now operate in a global environment, are driven by technology and need to contend with all demographics. These organizations collect data (very large amounts of data – Big Data), deal with and process information (information overload) to make decision. These organizations are now looking for e-Services that are smart, intelligent, and capable of analyzing and making fast decisions with big data to be able to survive in a global economy and fickle customers. E-Services have enabled retail organizations to reach global customers and increased sales/profit.

2. Technological convergence

Information Communications Technology (ICT) is the simplest example of technological convergence. Before information technology, information was either disseminated via print media (newspapers), audio media (radios) or audio and video (TV). To consume this information, an individual was required to pay for and access three different technologies on three different mediums or devices. Now, via the Internet, on a webpage, a consumer can access all three media at once with a single technology. Therefore, the print, audio, and voice and audio technologies have all converged on a single web page and accessed via a single device. With this convergence, the user has control in terms of when, for how long and how many times, the user can access this information, with the option of archiving it for future retrieval and reference. According to Goundar [5], "with this simple example, we can say that the integration of different technologies into one and the provision of them as a single service is technological convergence".

Similar statements were echoed in an essay written by Papadakis [6], for the International Telecommunications Union. He defines the term technological convergence as "a process by which telecommunications, information technology and the media, sectors that originally operated largely independent of one another, are growing together". He adds "technological convergence has both a technical and a functional side. The technical side refers to the ability of any infrastructure to transport any type of data, while functional side means the consumers may be able to integrate in a seamless way the functions of computation, entertainment, and voice in a unique device able to execute a multiplicity of tasks." "Technological convergence if appropriately managed can play an important role in national economic and social development of every nation. Governments can capitalize on the opportunity to stimulate market development and meet previous unmet society communication needs", [6].

In an organization, analog phone lines, fax machines, and other stand-alone office equipment have converged into a computer network connected via digital data lines [5]. Now, from one device, a number of different services on different technologies can be accessed. VoIP has replaced phone line communications and email attachments have replaced faxes and postal. "Technological convergence results in greater benefits from increased diversity in products and services in an organization. Using networking technology that connects all information and communications services with a single network, companies can add services to their previous ones, without new investments in infrastructure". The ability to integrate different technologies (technological convergence) seamlessly has resulted in disruptive technologies like Uber, Netflix, 3D Printing, Self-driving cars, drone deliveries, and the list is not exhaustive. Technological convergence emerged as a savior for all of us during the time of the COVID-19 pandemic.

Already converged technologies like Artificial Intelligence, Blockchains, Robotics, Bioinformatics, Data Science, are converging again to provide totally automated and intelligent services that did not exist before. The article "The Technological Convergence Innovation" authored by Adams et al. [7], "discusses the acceleration and Integration of Everything (AIE), i.e., of all forms of electronic devices into a distributed communications grid that will, inexorably, ubiquitously change the way we exist towards a convergent singularity of robotics, informatics, genetics, and nanotechnology. The changes may be more than the collective or individual human psyche is prepared to engage and will require that societies get used to these changes and incorporate them. The use of communication and information technology is also as important for sociology as it is for any other subject as it influences and is influenced by different kinds of policy, about citizenship" [7]. This research article indicates that technological convergence still has a long way to go and there will be further disruptions.

3. Digital transformations

We have entered the era of the Fourth Industrial Revolution (4IR). The Fourth Industrial Revolution is building on the Third (the digital revolution). The digital revolution has been occurring since computers came into existence. It is characterized by a fusion of technologies that is blurring the lines between the physical, digital, and biological spheres. The "digital revolution is transforming the world as we know it at unprecedented speed. Digital technologies have changed the way businesses operate, how people connect and exchange information, and how they interact with the public and private sectors". According to Goundar [8], "businesses and citizens alike need to be aware of and possess appropriate skills and infrastructures to capture the enormous value created by the digital economy and make a success of digital transformation".

The digital revolution enabling the Fourth Industrial Revolution that the world is witnessing. It will manifest a significant impact on how we live, work, and interact with people and machines. Organizations have no other alternatives, but to step into digital world for their survival and to sure sustainable competitive edge. The governments across the world came up with an initiative to amplify the scope of e-Governance by floating so many digital programmes and undergoing digital transformations. The objective of digital transformation is to provide the necessary infrastructure as a basic utility for every citizen of their country to provide on demand services with superior governance and empowering their citizens digitally.

Security and privacy are the major barriers in adopting the digital mechanisms. Organizations and individuals are concerned about their private and financial data. The security mechanisms that safeguard and monitors the sensitive information should be standardized and the security policies should be assessed on a regular basis. "Nowadays more and more innovative applications are using standardized cryptographic mechanisms to explore many new innovative digital financial applications and various decentralized applications that eliminate the need for third party intermediaries, such as identity management, credit management, distrusted ledger, crowdfunding, crowdsourcing, blockchains, P2P insurance, smart contracts, supply chain management, online voting, medical records, to name a few".

Digital transformation marks a radical rethinking of how an organization uses technology, people, and processes to fundamentally change business performance. "Digital transformation is a foundational change in how an organization delivers value to its customers". According to Negreiro and Madiega [9], "Internet and digital technologies are transforming our world. For decades, societies and economies have been experiencing a radical digital transformation, fostered by 'digitalization' and the speeding up of many kinds of interaction through the increasing number of connected devices and data flows".

Digital transformation covers both the integration of digital technologies by enterprises and the impact on society of new technologies, such as the Internet of Things (IoT), cloud computing, innovative digital platforms and blockchain technologies, writes [9] in their European Union Brief on Digital Transformations. "It is becoming an increasingly important condition for modern economies to thrive and has the potential to affect many sectors of the economy (including transport, energy, agri-food, telecommunications, financial services, factory production and health care) and to transform people's lives. According to the OECD, the greater computing power of consumer devices, which are available at ever more affordable prices, is accelerating this transformation". According to Goundar [8], in his book, the Impact of Digital Transformations on Security Policies and Standards, "Artificial Intelligence (AI) and Advanced Robotics are viewed as an important manifestation of the digital transformation.

3.1 The impact of digital transformations

As an example, the European Union plays an active role in shaping the digital economy, with cross-policy initiatives that range from boosting investment to reforming EU laws, to non-legislative actions to improve Member States' coordination and exchange of best practices. The 2014–2019 parliamentary term has seen a number of initiatives in the areas of digitalization of industry and public services, investment in digital infrastructure and services, research programmes, cybersecurity, e-commerce, copyright, and data protection legislation. There is a growing awareness among EU citizens that digital technologies play an important role in their everyday lives [9].

In a 2017 survey, two-thirds of Europeans said that these technologies have a positive impact on society, the economy, and their own lives. However, they also bring new challenges. A majority of respondents felt that the EU, Member States' authorities and companies need to take action to address the impacts of these technologies. The European Union will increase its support for digital transformation in the coming years, as illustrated by the recent proposal for the Digital Europe programme (for 2021–2027) – which would be the first ever funding programme dedicated solely to supporting digital transformation in the EU.

Negreiro and Madiega [9], states that further EU action will doubtless be needed, notably to increase infrastructure investment, boost innovation, foster digital champions and businesses digitalization, reduce existing digital divides, remove remaining barriers in the digital single market and ensure an adequate legal and regulatory framework in the areas of advanced computing and data, artificial intelligence, and cybersecurity. The European Parliament, as co-legislator, is closely involved in shaping the policy framework that will help citizens and businesses fully exploit the potential of digital technologies.

Digital transformation is a revolutionary technology that will play a vital role in most of the major industries. Apart from governance, it can be used for a wide variety of applications such as tracking, ownership, physical assets, voting rights, security and encryption of various digital resources and access of online, distributed platforms. Digital transformation encompasses some of the following indexing keywords: Applied Cryptography, Blockchain Technologies, Data Security and Protection, Digital Financial Applications, e-Services, Legal, Regulatory, and

Compliance Issues, Cyber-Physical Systems, Security Vulnerabilities, Security and Privacy Applications, and Trust Models. However, this is not an exhaustive list of indexing keywords. Organizations that have been transformed and undergone digital transformation find it easier to adapt to changes in economy, are competitive, innovative and deliver to the stakeholders the appropriate dividends. Digital Transformations are now being taught as course in universities and corporations.

Successful digital transformation requires an organization to have an embedded digital culture. As organizations advance from pilot digital transformation programs to wide scale adoption, they often run into unexpected obstacle: culture clash, laments Hemerling et al. [10]. According to the authors of the article "It's Not a Digital Transformation without Digital Culture", being a digital organization means not only having digital products, services, and customer interactions but also powering core operations with technology. Becoming one, therefore, requires a tectonic change in the activities employees perform as well as in their individual behaviors and the ways they interact with others inside and outside the organization. Leaders need to acknowledge digital transformation as the fundamental, strategic paradigm shift that it is. Like any major transformation, a digital transformation requires instilling a culture that supports the change while enabling the organizations overarching strategies.

4. The "e" of services

According to Hull et al. [11], "the emerging paradigm of electronic services (**e-Services**) promises to bring to distributed computation and services the flexibility that the web has brought to the sharing of documents". They state that "the last several years have seen an explosion of activity around electronic services, in e-commerce, in science, and in telecommunications". According to them, "the fundamental objective of e-services is clear: to have a collection of network-resident software services accessible via standardized protocols, whose functionality can be automatically discovered and integrated into applications or composed to form more complex services".

Stafford [12], on the other hand writes "e-services represent a business model whose time has most certainly come. They provide a time-saving functionality that busy consumers can use readily. There are so many people from so many technical and academic specialties within the e-services arena that the sky is the limit in terms of interpretation. From the product marketing perspective, an e-service could be any electronically enabled aspect of customer utility. Technologists naturally view e-services as Web-delivered software functionality, often characterized under the rubric of Web services. The marketing and IT fields both have claims on the emerging e-services paradigm".

4.1 e-Commerce

e-Commerce "stands for electronic commerce and pertains to trading in goods and services through the electronic medium" [13]. "Business to Business (B2B), Business to Consumer (B2C), Consumer to Consumer (C2C) and similar opportunity help consumer preferences and consumer markets developing electronic infrastructure for challenges of the future" states Gupta [13]. Accordingly, she adds "E-commerce has revolutionized business, changing the shape of competition with Internet, the computer communication network creating an e-commerce market place for consumers and business". In addition, she continues "with developments in the Internet and Web-based technologies, distinctions between traditional markets and the global electronic marketplace-such as business capital size, among others-are gradually being narrowed down. The low cost of the PC and the growing use of the Internet is one of reasons for that. There is a growing awareness among the business community about the opportunities offered by ecommerce". According to the eCommerceGuide.Com, "Ecommerce, or electronic commerce, refers to transactions conducted via the internet. Every time individuals and companies are buying or selling products and services online, they are engaging in ecommerce".

4.2 e-Business

While some use e-commerce and e-business interchangeably, they are distinct concepts, according to Gupta [13]. "In e-commerce, information, and communications technology (ICT) is used in inter-business or inter-organizational transactions (transactions between and among firms/organizations) and in business-to-consumer transactions (transactions between firms/organizations and individuals). In e-business, on the other hand, ICT is used to enhance one's business. It includes any process that a business organization (either a for-profit, governmental or a non-profit entity) conducts over a computer-mediated network". A more comprehensive definition of e-business is "The transformation of an organization's processes to deliver additional customer value through the application of technologies, philosophies and computing paradigm of the new economy." Gartner.Com, describes e-Business as "any process that a business organization conducts over a computer-mediated network. Business organizations include any for-profit, governmental, or nonprofit entity. Their processes include production-, customer-, and internal- or management-focused business processes".

4.3 e-Learning

What is the true model of e-Learning? According to Mayes and De Freitas [14], "It is arguable that there are really no models of e-learning per se – only e-enhancements of existing models of learning. Technology can play an important role in the achievement of learning outcomes, but it is not necessary to explain this enhancement with a special account of learning. Rather, the challenge is to describe how the technology allows underlying processes common to all learning to function effectively. A true model of e-learning would need to demonstrate on what new learning principles the added value of the 'e' was operating. Where, for example, the 'e' allows remote learners to interact with each other and with the representations of the subject matter in a form that could simply not be achieved for those learners without the technology, then we may have a genuine example of added value. However, in this example the role of the technology may be primarily to get remote learners into a position to learn as favorably as if they were campus-based, rather than offering a new learning method. In such a case the enhancement is an educational one, though the underlying learning theory explains both campus-based and distance learning with the same theoretical constructs".

4.4 e-Government

In search for a response to the question. "What is e-Government?", Fang [15] response is as follows "Governments worldwide are faced with the challenge of transformation and the need to reinvent government systems in order to deliver efficient and cost-effective services, information and knowledge through information and communication technologies. Development of Information and communication technologies catalyzed and led up to E-government". In his paper,

"E-government is defined as a way for governments to use the most innovative information and communication technologies, particularly web-based Internet applications, to provide citizens and businesses with more convenient access to government information and services, to improve the quality of the services and to provide greater opportunities to participate in democratic institutions and processes. E-government presents a tremendous impetus to move forward in the 21st century with higher quality, cost-effective, government services and a better relationship between citizens and government". According to the United Nations e-Government Knowledge Base, "E-government has been employed to mean everything from 'online government services" to 'exchange of information and services electronically with citizens, businesses, and other arms of government".

4.5 e-Health

According to Eysenbach [16], "Internet is opening up to the area of health care. Intel, for example, referred to e-health as a concerted effort undertaken by leaders in health care and hi-tech industries to fully harness the benefits available through convergence of the Internet and health care." "Because the Internet created new opportunities and challenges to the traditional health care information technology industry, the use of a new term to address these issues seemed appropriate." Eysenbach [16], add "these new challenges for the health care information technology industry were mainly (1) the capability of consumers to interact with their systems online (B2C = "business to consumer")"; "(2) improved possibilities for institution-to-institution transmissions of data (B2B = "business to business")"; "(3) new possibilities for peer-to-peer communication of consumers (C2C = "consumer to consumer")". So, how can we define e-health in the academic environment? Eysenbach [16] "e-health is an emerging field in the intersection of medical informatics, public health and business, referring to health services and information delivered or enhanced through the Internet and related technologies. In a broader sense, the term characterizes not only a technical development, but also a state-of-mind, a way of thinking, an attitude, and a commitment for networked, global thinking, to improve health care locally, regionally, and worldwide by using information and communication technology".

4.6 e-Services research and practice

The academic community has been quite active in regard to e-Services research and practices. A search on Google Scholar <https://scholar.google.com/> using the keywords "e-Services Research and Practice" produces 44, 200 results in 0.27 seconds, while a search using the same keywords on Google <https://www.google.com/> produces about 1,730,000,000 results in 0.85 seconds. Most of the research has been on topics such as:

- Creating e-Services using Open Government Data
- Conceptualization of e-services Quality and e-satisfaction
- How to Research e-Services as Social Interaction
- Trust in e-Services: Technologies, Practices and Challenges
- e-services as Resources in Customer Value Creation

- Exploring Consumer Evaluations of e-Services: A Portal Site
- Demand Driven Development of Public e-Services
- Trust and Risk in Consumer Acceptance of e-Services

As listed above, it is evident that the research community is actively finding ways to build trust in e-Services by minimizing risks. There is also indication that e-Services results in customer value creation, provides quality service and is demand driven.

For example, in the research paper titled "e-Services: A Synthesis and Research Agenda", researchers [17] focuses on the services marketing research. They state that "services marketing research increases in both intensity and relevance as services contribute an increasing share of the world's economy and as firms and their customers increasingly interact through electronic networks". According to the researchers, "e-services present sharp new challenges to both researchers and practitioners, because the processes from beginning to end of the e-service value chain are markedly different than those for offline services and because the electronic environment offers increased flexibility throughout the value chain". This flexibility, they add, "creates the requirement to impose some sort of structure on all of the possible service and channel design choices. e-Service flexibility creates an opportunity, and the need, to think about the consumer early in the design process". Finally, they claim "flexibility makes it difficult, but critically important, to consider various scenarios suggesting future developments in e-services. Organizations may add value to existing goods and services with complementary e-services. They provide an overview of the past and some projections for the future in the new field of research and practices in e-Services.

"How e-Services are Manifested in Practice" is a research paper published in 2017, where authors [18] have looked the actual practice of e-Services in the public sector. They state that e-Services practiced in the public sector are based various models and frameworks that are not relevant any more. Over time, there have been advancements, new developments and new e-Services solutions that need to be taken into consideration to come up with new e-Services practices. They carry out an empirical investigation on how e-Services can be manifested in practice based on qualitative interviews with employees involved with e-Service development. Based on their empirical investigation they conclude that "e-service can take on many different forms within an organization; ranging from downloadable forms, to complicated self-service systems that require expertise knowledge and IT-systems with specific processing capacity. The notion that all services mediated through a website can be understood under one general umbrella term, without further categorization, needs to be challenged". The practice of e-Services in the public and private sectors are marginally different. While the private sector e-Services like e-Commerce (B2C) and e-Business (B2B) are profit oriented, public sector e-Services like e-Government and e-Governance and services oriented (Government to Citizen, G2C).

5. Future trends of e-services

At the time of writing (June 2021), we are in the midst of COVID-19 pandemic. Citizens in most of the developed and developing countries are facing wave after wave of coronavirus infections and deaths. The following has been extracted from the World Health Organization's website "Globally, as of 4:12pm CEST, 25

June 2021, there have been 179,686,071 confirmed cases of COVID-19, including 3,899,172 deaths, reported to WHO. As of 23 June 2021, a total of 2,624,733,776 vaccine doses have been administered". Source: <https://covid19.who.int/>, Retrieved, 27 June 2021. According to the World Trade Organisation [4], "the enforcement of social distancing, lockdowns and other measures in response to the COVID-19 pandemic has led consumers to ramp up online shopping, social media use, internet telephony and teleconferencing, and streaming of videos and films". They add "governments have adopted new measures, and the private sector has also acted, to respond to and ensure that e-commerce can help to alleviate some of the challenges faced in combatting the virus. These have included increasing network capacity, offering expanded data services at little or no cost, lowering or scrapping transaction costs on digital payments and mobile money transfers, improving delivery services and other logistics, using digital tools to enforce measures and disseminate information, promoting telehealth services, and leveraging ICT for surveillance".

One of the positives, out of this coronavirus pandemic, is the increase in uptake of e-Services, especially, video conferencing services. With lockdowns and social distancing in place, e-Services (video conferencing) have provided a platform to enable us to work from home (remote work). The "reduced travel time and cost, the importance of involving employees in determining strategic goals, and the rising need for virtual meeting rooms have overall increased the spending of companies on video conferencing solutions", (e-Services). Organizations and their employees are using video conferencing platforms like Zoom, Skype, Microsoft Teams, Cisco's WebEx, GoToMeeting, and Google Meet, etc. to virtually meet and organize work, tasks, and activities. According to Richter [19], "Zoom saw its revenue skyrocket throughout the fiscal year ended January 31, 2021. Following a 169-percent revenue increase in the first quarter, year-over-year revenue growth accelerated to 355, 367 and 369 percent, respectively, in the second, third and fourth quarter. For the twelve months ended January 31, Zoom's revenue amounted to $2.65 billion, up from just $623 million the previous year". "The global video conferencing market size is expected to grow from USD 9.2 billion in 2021 to USD 22.5 billion by 2026, at a Compound Annual Growth Rate (CAGR) of 19.7% during the forecast period", (marketsandmarkets.com).

According to the World Trade Organisation [4], "e-commerce in services that can be delivered electronically has flourished, with demand rising sharply. One example is media services. Facebook reports that its online messaging, voice and video call services are up by more than 50 per cent, with Italy showing a 70 per cent surge overall, and a 1,000 per cent increase in group calls". New and innovative models of doing business are aplenty with e-Services platforms. For example, restaurants that cannot have customers eat-in, are taking takeaway and delivery orders from customers using whatever e-platform, social media platform, or messaging apps. Farmers are also receiving orders for their products and making deliveries. Entrepreneurs have found that with the use of one of the e-Services platforms, they will be able to continue their businesses. Mobile payments, and mobile payments are replacing credit cards with these small scale or solo business owners. Another major use of e-Services (e-Learning) has been in education. Most of the students are now studying online, attending Massive Open Online Courses (MOOC), as well as learning whatever they can from YouTube videos. Platforms such as Amazon have chipped in, "offering the public sector free access to its remote education, remote working and research tools, and Cisco has made its WebEx video conferencing tool free of charge".

Author details

Sam Goundar
The University of the South Pacific, Fiji

*Address all correspondence to: sam.goundar@gmail.com

References

[1] Lewis, P. H. (1994). Attention Shoppers: Internet is Open. Business Section. New York Times.

[2] Coppola, D. (2021). E-commerce worldwide - Statistics & Facts. Key Figures of e-Commerce – e-Commerce. https://www.statista.com/topics/871/online-shopping/ (Retrieved 26 June 2021).

[3] Sabanoglu, T. (2021). Amazon Prime Day - Statistics & Facts. B2C e-Commerce – e-Commerce. https://www.statista.com/topics/5385/amazon-prime-day/ (Retrieved 27 June 2021).

[4] World Trade Organisation. (2020). E-Commerce, Trade, and the COVID-19 Pandemic – Information Note. WTO Secretariat.

[5] Goundar, S. (2021). Introduction to Enterprise Systems and Technological Convergence. Enterprise Systems and Technological Convergence: Research and Practice, 1.

[6] Papadakis, S. (2007). Technological convergence: Opportunities and challenges. *Ensayos de la Unión Internacional de Telecomunicaciones*.

[7] Adams, T. L., Taricani, E., & Pitasi, A. (2018). The technological convergence innovation. International Review of Sociology, *28*(3), 403-418.

[8] Goundar, S. (2020). Introduction to the Impact of Digital Transformation on Security Policies and Standards. IGI Global.

[9] Negreiro, M., Madiega, T. (2019). Digital Transformations. Briefing – EU Policies – Delivering for Citizens. European Parliament Research Services. Members' Research Service PE 633.171.

[10] Hemerling, J., Kilmann, J., Danoesastro, M., Stutts, L., & Ahern, C. (2018). It's not a digital transformation without a digital culture. Boston Consulting Group. Retrieved December, *25*, 2018.

[11] Hull, R., Benedikt, M., Christophides, V., & Su, J. (2003, June). E-services: a look behind the curtain. In Proceedings of the twenty-second ACM SIGMOD-SIGACT-SIGART symposium on Principles of database systems (pp. 1-14).

[12] Stafford, T. F. (2003). E-services. Association for Computing Machinery. Communications of the ACM, 46(6), 26-26.

[13] Gupta, A. (2014). E-Commerce: Role of E-Commerce in today's business. International Journal of Computing and Corporate Research, 4(1), 1-8.

[14] Mayes, T., & De Freitas, S. (2007). Learning and e-learning. Rethinking pedagogy for a digital age, 13-25.

[15] Fang, Z. (2002). E-government in digital era: concept, practice, and development. International journal of the Computer, the Internet and management, 10(2), 1-22.

[16] Eysenbach, G. (2001). What is e-health? Journal of medical Internet research, 3(2), e20.

[17] Hofacker, C. F., Goldsmith, R. E., Bridges, E., & Swilley, E. (2007). E-services: a synthesis and research agenda. In E-Services (pp. 13-44). DUV.

[18] Lindgren, I., & Melin, U. (2017, September). Time to refuel the conceptual discussion on public e-services–revisiting how e-services are manifested in practice. In International

Conference on Electronic Government (pp. 92-101). Springer, Cham.

[19] Richter, F. (2021). Zoom Keeps Momentum as Workers Stay at Home. Zoom Video Communications, Mobile Communications. https://www.statista.com/chart/21906/zoom-revenue/ (Retrieved 27 June 2021).

Chapter 2

Consumer Perception of Electronic Commerce – Incorporating Trust and Risk with the Technology Acceptance Model

Sam Goundar, Kunal Lal, Ashmita Chand and Pranav Vyas

Abstract

The rapid growth in internet technology and the innovations in smart devices has had its impact on how businesses used to operate a decade ago to how consumers followed traditional shopping behavior when goods or services was required. The rise in internet users presents the opportunity of Electronic Commerce, the idea of providing and accessing goods and services using internet, such as selling goods and services, performing financial transactions and even booking an appointment. The two leading theories - perceived ease of use and usefulness have been employed in numerous information systems research studies to help understand consumer behavior in e-commerce environments. As the acceptance of e-commerce in Fiji is relatively low, this paper integrates the findings of previous researches with the perceived trust and perceived risk of consumers to provide a comprehensive framework of online shopping behavior based on the Technology Acceptance Model. To gain an in-depth understanding of consumers' acceptance of e-commerce an online questionnaire was used with questions which focused on these disciplines. The outcome of which verified that there were correlations between trust, perceived risk, perceived ease of use, perceived usefulness and usage of e-commerce. Through the use of this research, businesses in Fiji may incorporate features as per user expectations and have a good background before venturing into e-commerce.

Keywords: Consumer behavior, online shopping, perceived ease of use, perceived risk, perceived trust, perceived usefulness, technology acceptance, transaction intentions, e-commerce

1. Introduction

Tremendous growth has been experienced in ICT and much of the advancement may be attributed to the various types of ICT ventures that invent and commercialize a variety of ICT products [1]. With emerging information technology, more and more people can access internet connectivity everyday using smart devices. According to Internet World Stats, as of December, 2017 54.4% of the world population use the internet [2].

Many firms approach the e-commerce market with no clear idea of where they are going. In this emerging global trend of business, technology is increasingly becoming a necessary component and a strong promoter for electronic sales. The use of internet in business has enhanced productivity, encouraged greater customer participation and satisfaction, besides reducing costs. With the current developments in internet and web-based technologies, distinctions between traditional markets and the global electronic marketplace-such as business capital size, among others-are gradually being narrowed down. One of its utmost result is the convenience of online shopping. Online shopping can be considered as an exchange of money, time and effort for getting products and services.

e-Commerce is beginning to gain traction in the Pacific Island Countries (PICs) [3–5] as the affordability and availability of Internet connectivity becomes better. Many businesses in sales and retail are trying to take advantage of e-Commerce's many capabilities, flexibilities for service provision, convenience, and cost savings [6, 7]. A few researches [3–5] have looked into the availability, provision, and challenges of e-Commerce in the region from the business owner's perspective. Consumers perception of e-Commerce is lacking or non-existent for Pacific Island Countries. Preliminary investigation with many consumers found trust and risk as being a common inhibitor to engage in e-Commerce and the same has been found in published literature [8–10]. This research study aims to fill the research gap on consumers' perception of e-Commerce taking into consideration the trust and risk factor. Based on this knowledge, the relevant stakeholders, namely businesses, consumers and government as regulators can create a conducive e-Commerce landscape.

The Technology Acceptance Model (TAM) has been used to ground this research from a theoretical perspective and align it with a validated research model. The Technology Acceptance Model (TAM) [11] has been widely used by researchers to check and test the acceptance of e-Commerce technology by users [3, 12–15]. The model determines how users might react to the acceptance of new technology based on their Perceived-Ease-Of-Use (PEOU) and Perceived Usefulness (PU) of technology [3, 11]. e-Commerce being a new technology for the consumers in the Pacific Island Countries (PICs), and its acceptance and use could easily be determined with the Technology Acceptance Model. Furthermore, the TAM is versatile enough to be easily customized and extended to incorporate trust (Perceived Trust) and risk (perceived Risk) factors for consideration into use and acceptance of e-Commerce. Therefore, the Technology Acceptance Model that could be customized and expanded to incorporate trust and risk was preferred over other models such was Unified Theory of Acceptance and Use of Technology (UTAUT) [16], Theory of Planned Behavior (TPB) [17], and Theory of Reasoned Action (TRA) [18].

In this paper, perceived trust, perceived risk, perceived usefulness and ease of use is integrated using the Technology Acceptance Method (TAM) to explore its effect on customers' acceptance of e-commerce in Fiji. Data was gathered from 82 respondents by use of online questionnaire, links of which were distributed using means of social media, emails and forums. It is empirical to maintain the consistency of perceived trust and perceived risk with perceived ease of use and perceived usefulness, which is further used to measure general perceptions.

This research study and the results of this research intends to add a body of knowledge on consumers perception of e-Commerce that is lacking or non-existent for Pacific Island Countries. The implications of filling this research gap with the body of new knowledge are many for the society. For example, e-Commerce now contributes substantially to a countries Gross Domestic Product (GDP) [19, 20]; creates employment [21, 22]; businesses can saves costs on all the physical resources required to operate a brick and mortar retail outlet [21, 22]; customers can get cheaper products with the removal of the middle-man [21, 22]. Within the Pacific

Island Countries, e-Commerce is seen as a pathway for Small and Medium Enterprises (SME) and sole entrepreneurs to get into the market without requiring the upfront capital to establish a physical presence. But venturing into any business, including e-Commerce is a risk. Understanding consumers' perception of e-Commerce, their perceived ease of use and usefulness, building trust and mitigating risks will go a long way in ensuring that consumers continue engaging in e-Commerce.

In a survey conducted by the Reserve Bank of Fiji, it was discovered that 88.8% of the respondents preferred to use cash over electronic money [23]. This meant that even though electronic money is available, there exists a high level of hesitancy in consumers. Should the hesitancy persist, Fiji may not be able to fully capitalize in the e-commerce sector. Hence the need for research to investigate and evaluate the factors associated to with a consumers' trust and their perceived risk towards engaging in e-commerce.

2. Literature review

Technology Acceptance model was used to gain knowledge of its users towards accepting any new information technology thought is four key variables; Perceived Trust, Perceived Risk, Perceived Usefulness and Perceived Ease-of-Use. The Concept of Perceived Trust is the key as every aspect of personal interaction with information technology is based on trust in one way or another [24]. The users of the e-commerce platform will need to invest their trust into the system to safeguard their personal information such has age, name, location, bank card numbers and pin codes. The users will have to have a level of reliability on the information system that the personal information entered will remain confidential and trust the merchant offering goods and services through e-commerce is not fraudulent.

Perceived Risk is a key influential factor for Perceived Trust. Perceived risks the uncertainty the consumer has while deciding to transact on an e-commerce website. The term perceived risk means the individual's subjective belief about potentially negative consequences from his/her decision [25]. Certainly, people may experience a certain degree of risk when purchasing a product through a web-based shopping channels because perceived risk has the characteristics to increase vulnerabilities and generate inhibiting aspects to consumers' willingness to participate in online shopping activity [26]. The consumer is faced with financial risk, functional risk, reputation risk and time risk. These risk are beyond a consumers control and the consumer may fear falling victim to a fraudulent activity.

Perceived Usefulness is defined as the degree to which a person believes that using a particular system would enhance his or her job performance [27]. The consumer evaluates the usefulness of using the e-commerce website to purchase goods and service again the traditional walk in the shop practice. Will the use of e-commerce site save him or her time from visiting the shop, is the price of the product cheaper if purchased online directly from supplier, and is there more variety of goods of service available online then local vendors. These are few of the assessing questions a consumer asks themselves before deciding on the Usefulness of the e-commerce websites.

Perceived Ease of Use refers to the degree to which a person believes that using a particular system would be free of effort [27]. The e-commerce website design need to factor in the internet connectivity of the country such as Fiji so that the pages are not overloaded with graphic which would take time to load in slow internet speeds leading to consumers losing interest. The e-commerce site needs to be user friendly, the web-content should be precise yet simple for its customers to easily understand.

E-commerce webpages should not feel cumbersome to online visitors, providing easy simple navigation steps to purchase goods and services without spending much time and clicks on multiple links. If a system is relatively easy to use, individuals will be more willing to learn about its features and finally intend to continue using it [28].

Numerous businesses are unable to turn e-commerce into an opportunity which could be used for the benefit of both businesses and consumers. Many new enterprises in developing countries lack the technical expertise and knowledge on the power of using internet as a tool to boost the sale of goods and expand the reach of services offered. The challenge of convincing decision makers of companies to invest into e-commerce for long term profitability is a hindrance caused by lack of knowledge in the technological advancements of this era. A computer-interface to human relationship is evident in e-commerce which results in risk being a major and common factor which repels consumers from e-commerce. A narrow definition provided in [29] states that e-commerce is the "sale or purchase of goods or services, whether between businesses, households, individuals, governments, and other public or private organizations, conducted over the Internet" where by ordering of goods and services are done online but the final delivery of the good or service may be done online or offline.

Businesses that wish to gain consumers' loyalty and construct long term associations must research on factors which may affect how a consumer reacts to their e-commerce site. Factors such as; consumer's expertise on use the of smart devices and computers, knowledge about internet and accessibility, the type of goods and services which appeal to consumers using e-commerce, the target audience for e-commerce goods and services, consumer financial constrain to invest in internet verse the e-commerce gain and most importantly the widespread age group of the country. A country with majority of population born and living in the period of internet advancement would be more loyal for e-commerce services. An example of gaining consumers' trust is proposed via a model in [30] whereby with the use of the model consumers can avoid deception in transactions by comparing new transactions with old ones. A study by Jones and Leonard [31] revealed that the trust of consumers in e-commerce is heavily influenced by internet safety. Together with the ease of use of e-commerce interface, as explained in [32] aspects such as diversity, discounts, convenience and influence via social networks also play a vital role in driving consumers towards the use of e-commerce.

Preceding e-commerce research had the absence of explaining the perceived risk, perceived usefulness, perceived trust and perceived ease of use from a consumer's approach. The research gap in the confidence of consumer to engage in the use of local vendor e-commerce site in evaluation to foreign e-commerce site, evolution of risk for after sale support for e-commerce vendor local and overseas, the usefulness of adapting to e-commerce verse the reality of beneficial gain it would provide consumers of developing country such as Fiji. Consumers and other stakeholders are often cautious of transacting online. Not only are they making an online transaction in exchange of electronic cash but are also pushing confidential information in the hands on the business. The details provided by consumers has a financial value though it may not be seen at beginning. As explained in [33] "e-business owners are exploiting the user's privacy for the growth of their business". Electronic commerce businesses buy information about individuals, their personal details, shopping habits and web page visitation listings [34]. Major ethical issues companies face when doing business related to e-commerce are privacy, security, trust, intellectual property rights, some environmental issues. In order to manage ethical issues in e-commerce, managers have to help the company to design an effective strategy and long-term management procedure as well as security policies and training. Different companies and different fields of e-commerce have different security problems, thus, different strategies, "however it must be legal, feasible, effective and innovative" [35].

3. Research model and hypothesis

Various factors affect a consumer's intention to transact online such as perceived risk, trust in the online store, perceived reputation and size of the business [36]. The Technology Acceptance Model (TAM), which is used as the base for this research is extended to comprise of specific concerns mainly perceived trust and perceived risk; integration of which make for the core objectives of this paper. Similar context is claimed in [37] that the primary concern of the customers in e-shopping context is trust and perceived ease of use. This research attempts to construct a model for analyzing e-commerce adoption from a consumer's perspective based on TAM, which has been widely studied and accepted as a powerful framework for researching IT adoption and usage.

This leads to suggest the following hypotheses:

3.1 Perceived risk [PR]

It refers to the degree of consumer distrust that the use of e-commerce is unsafe. Whether it be of consumers' privacy and safety or the risk involved with the processing of transactions. Perceived risk consists of information misuse, failure to gain product benefit and functionality inefficiency risk [38]. When consumers' buy from web, they provide their personal and financial information, both of which revealed on network might be misused. Also, during web purchasing the consumer may face problems such as the purchased product not meeting expectations because the customer does not get an opportunity to inspect before buying. Another problem is the delay in delivery. Regarding this paper, consumers who believe that the level of perceived risk is high in terms of risks to personal information and risks in transaction processing; do not tend to adopt e-commerce. Keeping this observation, the following is hypothesized:

H1: Perceived risk (PR) negatively influences the consumer's intention to transact online.

H2: Perceived risk of information misuse negatively influences perceived ease-of-use (PEOU).

H3: Perceived risk negatively influences perceived usefulness (PU).

3.2 Perceived trust [PT]

For vendors to excel in e-commerce, online trust is vital as it is claimed that in the virtual world the issue of trust gets enlarged [39]. A lack of trust discourages online consumers from participating in e-commerce and deters any purchase over the internet. Online trust is one of the factors that is frequently associated with the failure or success of online ventures and its multi-dimensional character makes it a complicated issue. On the assumption of trust's relation with purchasing online the following is hypothesized:

H4: Trust in e-commerce positively influences the perceived risk associated with transacting online.

H5: Trust in e-commerce positively influences the consumer's intention to transact online.

H6: Trust in e-commerce positively influences perceived ease-of-use (PEOU).

H7: Trust in e-commerce positively influences perceived usefulness (PU).

3.3 Perceived usefulness [PU]

Is defined as the degree to which a consumer believes that using a system would enhance their job performance [40] and how effectively it improves the way a consumer can complete a task. Therefore, accurate information should be provided in conjunction with the core features of the e-commerce site. Based on these findings, the following hypotheses are proposed:

H8: Perceived usefulness positively influences the consumer's intention to transact online.

3.4 Perceived ease of use [PEOU]

Ease of use is defined as the consumers' perception that use of the new technology will be free of hardship and low in complexity. This is also to allow consumers to experiment with new innovations and evaluate its benefits easily and at no extra cost. Out of the five principles for ease of use explained in [41], this paper focusses on the learnability of an e-commerce site; that is "how easy is it for consumers to accomplish basic tasks the first time they encounter the design". Applying this to the research, ease of use is the consumer's perception that shopping on the internet will involve only minimum effort. Similarly, a quota sample was used as the research sample in [42] which was then evaluated against the TAM. The research concluded with the result that consumers in Croatia, which is a developing country feel that getting information from e-commerce would be free of effort (ease of use) and useful in their retailing experiences. By applying these into e-commerce context, the following is hypothesized:

H9: Perceived ease-of-use positively influences perceived usefulness (PU).

H10: Perceived ease-of-use positively influences the consumer's intention to transact online.

4. Research model

The hypotheses discussed above are depicted in the proposed research model in **Figure 1**; where the correlations between each factor is shown. A plus sign depicts a

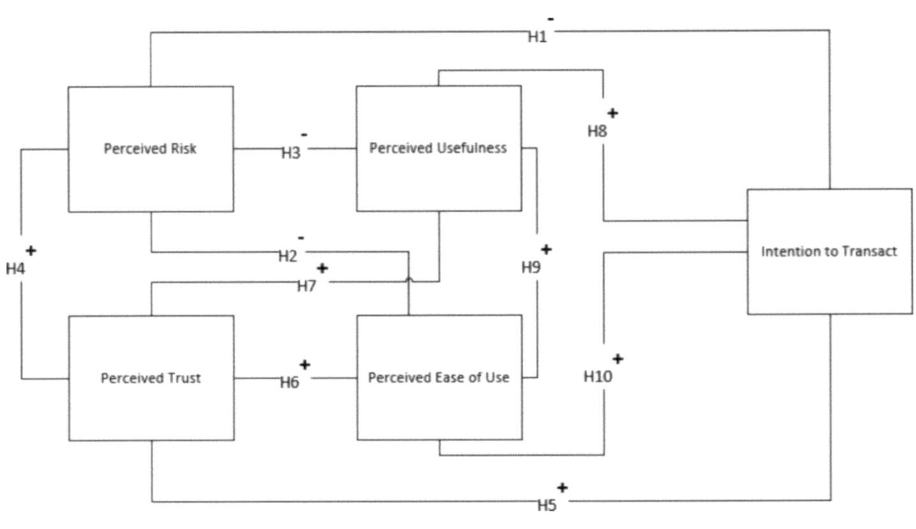

Figure 1.
Research model.

positive influence while a minus sign depicts a negative influence. For instance, perceived risk negatively influences a consumer's intention to transact (H1) while perceived trust positively influences a consumer's intention to transact (H5).

5. Theoretical background

Technology Acceptance Model (TAM): As explained by Venkatesh and Davis in [43] TAM is a model through which a consumer's intention to use a system is determined by two principles; perceived ease of use and perceived usefulness. This model has been successfully adapted and extended in developing countries like Jordan to aid in explaining online purchasing and consumers' behavior towards it. The simplicity and adaptability of this model is what makes it so popular and successful as it is easy to understand and permits to differentiate among the results for various information systems [44].

In context of perceived usefulness, a consumer may be targeting to achieve various advantages from e-commerce such as obtaining discounts, saving time and money, getting a wide variety of products to choose from, however this paper focusses on the type of information provided by e-commerce sites, where a consumer sees the information provided as useful or not. Perceived ease of use is simply described as when a consumer is easily able to perform the required tasks on an e-commerce site, such as searching for information, saving an order for a product and using the customer services feature. According to a research [45] customers who think that purchasing over internet is useful, and they can do it easily as they tend to adopt e-commerce.

6. Methodology

The objectives of this paper are reached through three methods. Firstly, by reading relevant literature already published in this context. The research done in the TAM model revealed that perceived trust and perceived risk was a made-up behavior of people who would make decision through observing other people's e-commerce experiences. The decision on perceived trust and perceived risk depended on the outcome of someone's e-commerce experience to be positive or negative. Weather the consumer was a victim of fraud and experienced financial loss or the consumer received a bargain on the selected good or service. This pre0-deciding factor was to learn from others experience and could impact on the future of the e-commerce by researchers, as perceived trust and perceived risk are subjective approach to e-commerce growth. The opportunity to execute studies with various designs is offered by means of conducting surveys whereby specific research questions can be answered [46]. With the aim of exploring and bringing out consumers' views on the theoretical concepts of TAM with trust and perceived risk, a quantitative research was conducted by the use of an online questionnaire. The questionnaire was designed to gain an insight on the confidence of a user towards the use of an online e-commerce site in comparison to their opinions on the perceived trust and perceived risk, and the frequency of the user's online e-commerce activity. Moreover, the decision to adopt a quantitative research method is supported by the fact that this method is useful in situations where data is analyzed and reported through statistical inferences and analyses [47]. The questionnaire was distributed to diversity group of people to collect as much sample data as possible. The diversity included different age groups, young teenagers to senior citizens, low to high computers literacy group and a variety of internet accessibility areas. A consumer's actual usage of e-commerce was used as the dependent variable as that was the main variable which was to be measured. Similarly, trust, perceived risk, perceived ease of use and perceived usefulness were used as independent variables which were claimed to be affecting the dependent variable. The survey was published online from 12th September 2018 to 23rd October 2018 and the link was circulated to as many consumers as possible via email and posts on social media accounts after which a total of 83 responses were collected. The primary participants were friends, families and colleagues of the researchers who after completing the survey further circulated the survey link. Lastly, by analyzing the data collected and using the Technology Acceptance Model (TAM) to validate the research questions. Earlier results showed that there are dependent variables such as age and computer literacy provided mixed results on an individual's willingness to overlook the perceive risk and trust against the benefits of e-commerce. Independent variables in our quantitative research was the locality of our consumer's and the accessibility to internet. The geographical landscape and internet still new to a country such as Fiji, the awareness of e-commerce website to its population and the changing idea of perceived trust and perceived risk a challenge.

7. Results

Validity is an extent to which a measure or set of measures correctly represents the concept of the study [48]. It is concerned with how well the concept is defined by the measures. While reliability is an extent to which variable or set of variables is consistent in what it is intended to measure [48].

The main difference with validity is, reliability relates to what should be measured not how it is measured [48]. **Table 1** below shows the reliability item (validity of each indicator).

Factor loadings value should be at least 0.5 to be considered acceptable [49]. In **Table 1**, all latent variables and hypothesis variables are above the acceptance range and will be included in the analysis.

The values suggests that with each variables or indicators are validating their construct (hypothesis). Using "PR" latent variable as example, the construct variable has 3 hypothesis (H1, H2, an H3) and all of them have loading values greater than 0.5.

Latent variable	Hypothesis variable	Standardized loadings	Critical ratio
PR	H1	0.724	13.581
	H2	0.751	19.924
	H3	0.756	15.052
PT	H4	0.730	11.556
	H5	0.767	20.626
	H6	0.813	22.512
	H7	0.781	18.646
PU	H8	0.797	16.832
PEOU	H9	0.767	15.917
	H10	0.761	19.310

Table 1.
Results of reliability tests.

Variable test	Question(s)
Trust	Do you trust online trade either locally or internationally?
Risk	Do you believe e-commerce websites have security features to carry out online financial transactions securely?
	How secure do you feel in sharing your personal details on an e-commerce website? (1 - least secure & 5 - most secure)
Usefulness	Do you believe e-commerce websites provide accurate information on products on sale and delivery times?
Ease of Use	Do you think e-commerce websites are user friendly?
Usage	Overall, what would your final choice be about e-commerce locally?

Table 2.
Research questionnaire.

Age group	Percentage
Under 21	0%
21–24	13.41%
25–34	58.54%
35–44	25.61%
45–54	0%
55–64	0%
65 or older	2.44%

Table 3.
Respondents' demographics by age group.

This indicates that those 3 indicators have a good validity level and they are to be considered as valid indicators to measure the PR (Perceived Risk) construct.

The core motive of the research was to capture consumer perception on trust, risk, usefulness, ease-of-use and usage. The design of the questionnaire questions captured just that. **Table 2** lists the key question(s) to measure and address each perception area.

The demographic details of respondents to the survey of 82 sample size as per **Table 3** shows that 97.56% of the respondents were within the age of 21–44.

The objectives of the paper as well as the analysis and results refer to the two main pillars of the TAM model; perceived ease of use and perceived usefulness. Following are statistical analysis and short discussions on each of the hypothesis of this paper. Since the sample size consisted of 82 respondents, and the Kolmogorov–Smirnov test was used to prove that the data was normally distributed, and all data types were ordinal; non-parametric spearman's correlational tests were conducted on the hypothesis.

7.1 H1: perceived risk [PR] negatively influences the consumer's intention to transact online

As depicted in **Table 4** a correlation was discovered between a consumer's perceived risk in transaction processing and intention to transact online ($r = -0.139$, p-value = 0.212). Since the correlation is negative and p-value significant, this implies that there is a significant relationship between the two; as a

Correlations				
			Risk transaction_ processing	**Intention_to_transact_ online**
Spearman's rho	Risk Transaction_ Processing	Correlation Coefficient	1.000	−0.139
		Sig. (2-tailed)		0.212

Table 4.
Correlation – PR in transaction processing & intention to transact.

consumer's perceived risk in transaction processing increases, the intention to transact online decreases.

As shown in **Table 5** a correlation was discovered between a consumer's perceived risk of personal information and intention to transact online (r = −0.022, p-value = 0.844). Since the correlation is negative and p-value significant, this implies that there is a significant relationship between the two that as a consumer's perceived risk of personal information increases, the intention to transact online decreases.

7.2 H2: perceived risk [PR] negatively influences perceived ease-of-use [PEOU]

As per the data presented in **Table 6** a correlation was discovered between a consumer's perceived risk in transaction processing and perceived ease of use (r = −0.264, p-value = 0.017). This implies that there is a relationship between the two that as a consumer's perceived risk in transaction processing increases, the perceived ease of use decreases. However, the probability for this correlation to arise due to sampling variation is low (that is, there is a correlation).

As shown in **Table 7** a correlation was discovered between a consumer's perceived risk of personal information and perceived ease of use (r = −0.177,

Correlations				
			Risk personal_ information	**Intention_to_transact_ online**
Spearman's rho	Risk Personal_ Information	Correlation Coefficient	1.000	−0.022
		Sig. (2-tailed)		0.844

Table 5.
Correlation - PR of personal information & intention to transact.

Correlations				
			Risk transaction_ processing	**Ease_of_use**
Spearman's rho	Risk Transaction_ Processing	Correlation Coefficient	1.000	−0.264*
		Sig. (2-tailed)		0.017

Correlation is significant at the 0.05 level (2-tailed).

Table 6.
Correlation - PR in transaction processing & PEOU.

Correlations				
			Risk personal_ information	**Ease_of_use**
Spearman's rho	Risk Personal_ Information	Correlation Coefficient	1.000	−0.177
		Sig. (2-tailed)		0.112

Table 7.
Correlation - PR of personal information & PEOU.

p-value = 0.112). This implies that there is a significant relationship between the two that as a consumer's perceived risk of personal information increases, the perceived ease of use decreases.

7.3 H3: perceived risk [PR] negatively influences perceived usefulness [PU]

A very strong correlation was discovered between a consumer's perceived risk in transaction processing and perceived usefulness (r = −0.306, p-value = 0.005) as shown in **Table 8**. This implies that there is a relationship between the two that as a consumer's perceived risk in transaction processing increases, the perceived usefulness of e-commerce decreases. However, the probability for this correlation to arise due to sampling variation is low (that is, there is a correlation).

As shown in **Table 9** a strong correlation was discovered between a consumer's perceived risk of personal information and perceived usefulness (r = −0.267, p-value = 0.015). This implies that there is a relationship between the two that as a consumer's perceived risk of personal information increases, the perceived usefulness of e-commerce decreases. However, the probability for this correlation to arise due to sampling variation is low (that is, there is a correlation).

Correlations				
			Risk transaction_ processing	**Usefulness**
Spearman's rho	Risk Transaction_ Processing	Correlation Coefficient	1.000	−0.306*
		Sig. (2-tailed)		0.005

**Correlation is significant at the 0.01 level (2-tailed).*

Table 8.
Correlation - PR in transaction processing & PU.

Correlations				
			Risk personal_ information	**Usefulness**
Spearman's rho	Risk Personal_ Information	Correlation Coefficient	1.000	−0.267*
		Sig. (2-tailed)		0.015

**Correlation is significant at the 0.05 level (2-tailed).*

Table 9.
Correlation - PR of personal information & PU.

7.4 H4: trust in e-commerce positively influences the perceived risk associated with transacting online

As depicted in **Table 10** a correlation was discovered between a consumer's trust in e-commerce and perceived risk in transaction processing (r = −0.188, p-value- = 0.092). This implies that there is a significant relationship between the two that as a consumer's trust in e-commerce increases, the perceived risk in transaction processing decreases.

As shown in **Table 11** a correlation was discovered between a consumer's trust in e-commerce and perceived risk of personal information (r = −0.246, p-value- = 0.026). This implies that there is a relationship between the two that as a consumer's trust in e-commerce increases, the perceived risk of personal information decreases. However, the probability for this correlation to arise due to sampling variation is low (that is, there is a correlation).

7.5 H5: trust in e-commerce positively influences the consumer's intention to transact online

As depicted in **Table 12** there exists a correlation between trust and a consumer's intention to transact online (r = 0.141, p-value = 0.206), thus concluding that as a consumer's trust in e-commerce improves, the consumer's intention to transact online also increases and vice versa.

Correlations				
			Trust	**Risk_transaction_ processing**
Spearman's rho	Trust	Correlation Coefficient	1.000	−0.188*
		Sig. (2-tailed)		0.092

**Correlation is significant at the 0.05 level (2-tailed).*

Table 10.
Correlation - trust & PR in transaction processing.

Correlations				
			Trust	**Risk_Personal_ Information**
Spearman's rho	Trust	Correlation Coefficient	1.000	−0.246*
		Sig. (2-tailed)		0.026

**Correlation is significant at the 0.05 level (2-tailed).*

Table 11.
Correlation - trust & PR of personal information.

Correlations				
			Trust	**Intention_to_transact_ online**
Spearman's rho	Trust	Correlation Coefficient	1.000	0.141
		Sig. (2-tailed)		0.206

Table 12.
Correlation - trust & intention to transact.

7.6 H6: trust in e-commerce positively influences perceived ease-of-use [PEOU]

As shown in **Table 13** trust and the consumer's perceived ease-of-use revealed a correlation (r = 0.159, p = 0.154). Hence implying that as trust in e-commerce improves so does the consumer's perception of ease of use and vice versa.

7.7 H7: trust in e-commerce positively influences perceived usefulness [PU]

As shown in **Table 14** trust and the consumer's perceived usefulness revealed a correlation (r = 0.322, p = 0.003) indicating that as trust in e-commerce improves, so does the consumer's perceived usefulness of e-commerce; and vice versa. However, the probability for this correlation to arise due to sampling variation is low (that is, there is a correlation).

7.8 H8: perceived ease-of-use positively influences perceived usefulness [PU]

As depicted in **Table 15** perceived ease of use and the consumer's perceived usefulness revealed a correlation (r = 0.174, p = 0.118). Hence implying that that as a consumer's perceived ease of use improves, so does the perceived usefulness of e-commerce and vice versa.

Correlations				
			Trust	**Ease_of_use**
Spearman's rho	Trust	Correlation Coefficient	1.000	0.159
		Sig. (2-tailed)		0.154

Table 13.
Correlation - trust & PEOU.

Correlations				
			Trust	**Usefulness**
Spearman's rho	Trust	Correlation Coefficient	1.000	0.322*
		Sig. (2-tailed)		0.003

**Correlation is significant at the 0.01 level (2-tailed).*

Table 14.
Correlation - trust & PU.

Correlations				
			Ease_of_use	**Usefulness**
Spearman's rho	Ease_Of_Use	Correlation Coefficient	1.000	0.174
		Sig. (2-tailed)		0.118

Table 15.
Correlation - PEOU & PU.

7.9 H9: perceived ease-of-use positively influences the consumer's intention to transact online

As presented in **Table 16** a correlation was discovered between ease of use and intention to transact online ($r = -0.015$, $p = 0.897$). Since the correlation is negative and p-value is more than zero, this implies that there is a relationship between the two that as a consumer's perceived ease of use increases, the intention to transact online decreases. This could be due the factor of trust, that if an e-commerce site is easy to use, then there may be something suspicious about it.

7.10 H10: perceived usefulness positively influences the consumer's intention to transact online

As shown in **Table 17** there exists a positive correlation between perceived usefulness and a consumer's intention to transact online ($r = 0.112$, p-value = 0.319), thus concluding that as consumer's perceived usefulness of e-commerce improves, the consumer's intention to transact online also increases and that there is a significant relationship between the two; and vice versa.

Spearman's Rho is a versatile measure of association tool that is being widely used [50, 51]. Many published research papers [52, 53] have used Spearman's Rho to measure the association and significance of risks and trust of using e-Commerce web sites. Spearman's Rho is a non-parametric test used to measure the strength of association between two variables, where the value $r = 1$ means a perfect positive correlation and the value $r = -1$ means a perfect negative correlation. Spearman's Rho is a nonparametric measure of rank correlation. It assesses how well the relationship between two variables can be described using a monotonic function. The hypothesis is already evaluating the impact of one concept on another. For example, H4: Trust in e-commerce positively influences the perceived risk associated with transacting online. Based on the aims of our research study to find the consumer perceptions of e-Commerce use and acceptance, Spearman's Rho is used to measure the strength of association between the Perceived Usefulness, Perceived Ease of Use, Perceived Risk and Perceived Trust. This association of positive or negative correlation is indicative of the impact of one concept on another.

The p-value tells you whether the correlation coefficient is significantly different from 0. (A coefficient of 0 indicates that there is no linear relationship.) If the

Correlations				
			Ease_of_use	Intention_to_transact_ online
Spearman's rho	Ease_Of_Use	Correlation Coefficient	1.000	−0.015
		Sig. (2-tailed)		0.897

Table 16.
Correlation - PEOU & intention to transact.

Correlations				
			Ease_of_use	Intention_to_transact_ online
Spearman's rho	Usefulness	Correlation Coefficient	1.000	0.112
		Sig. (2-tailed)		0.319

Table 17.
Correlation - PU & intention to transact.

p-value is less than or equal to the significance level, then you can conclude that the correlation is different from 0. The P-value is the probability that you would have found the current result if the correlation coefficient were in fact zero (null hypothesis). If this probability is lower than the conventional 5% ($P < 0.05$) the correlation coefficient is called statistically significant.

In most cases, in our results under 7.1 to 7.10, all the P values are more than 0, which indicates a correlation. In same cases, P values are less than 0.05, which still indicates a correlation but not that significant. Where the P values are less than 0.05, we have clearly explained the reasons.

8. Discussion

The objectives of this study were to explore the effect of perceived trust and perceived risk on the two dimensions of the TAM as well as on a consumer's final intention to transact. The **Table 18**, summarizes the results of the hypotheses for this paper.

As depicted in the **Table 18**, only one hypothesis (H10) was rejected, that is a direct relationship was expected between perceived ease of use and a consumer's intention to transact, however the correlation showed an inverse relationship implying that in Fiji, according to the sample and results, the easier an e-commerce site is to use, the more suspicious consumers may be of it, thus leading to a decline in its use.

The other hypotheses were accepted with relationship types as listed, for example;

- Inverse – as a consumer's perceived risk of e-commerce decreases, the intention to transact increases.

- Direct – as a consumer's trust in e-commerce improves, the intention to transact also increases.

The results of this research exhibit that both perceived risk and perceived trust affect a consumer's intention to transact as well as the ease of use and usefulness of an e-commerce site. Indicating that:

- Perceived risk negatively influences intention to transact, ease of use and usefulness of an e-commerce site. With this result, there were similarities and

Hypothesis	Correlation	Accept\ reject	Relationship type
H1	Perceived Risk ➔ Intention to Transact	Accept	Inverse
H2	Perceived Risk ➔ Perceived Ease of Use	Accept	Inverse
H3	Perceived Risk ➔ Perceived Usefulness	Accept	Inverse
H4	Trust ➔ Perceived Risk	Accept	Inverse
H5	Trust ➔ Intention to Transact	Accept	Direct
H6	Trust ➔ Perceived Ease of Use	Accept	Direct
H7	Trust ➔ Perceived Usefulness	Accept	Direct
H8	Perceived Usefulness ➔ Intention to Transact	Accept	Direct
H9	Perceived Ease of Use ➔ Perceived Usefulness	Accept	Direct
H10	Perceived Ease of Use ➔ Intention to Transact	Reject	Inverse

Table 18.
Hypothesis correlation relationship.

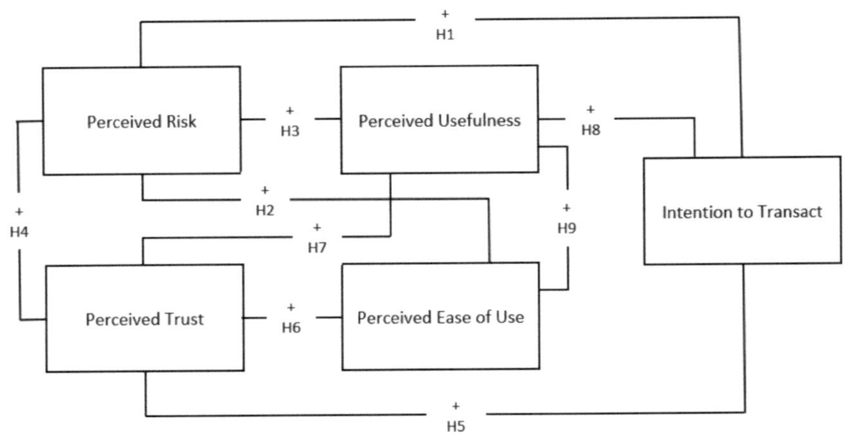

Figure 2.
Updated research outcomes model.

differences with the study [54] in a developing country Jordan, where by a correlation was found between perceived risk and intention to use (negative influence), however no significant relation found between perceived risk and ease of use and usefulness.

- Perceived trust positively influences intention to transact, ease of use and usefulness of an e-commerce site. Similarly, [55] proves that "trust increases the perceived usefulness of the website" as well as ease of use.

- Perceived usefulness positively influences intention to transact, which was also upheld in [56] as the correlation between both usefulness and actual use had a value of r = 0.732 confirming the relationship.

- Perceived ease of use positively influences perceived usefulness as was also proven in [57] via a study conducted on 155 online consumers.

Based on the results of the study, we have revised our research model from **Figure 1** above to that of **Figure 2** below. The updated research outcomes model in **Figure 2** below takes into consideration H10 that was rejected (that is a direct relationship was expected between perceived ease of use and a consumer's intention to transact). The new research outcomes model is based on **Table 18**: Hypothesis Correlation Relationship.

The study also revealed that consumers in Fiji have a good understanding of e-commerce as 70.73% of the respondents have used e-commerce to purchase goods. Also, e-commerce has made an impact on the consumers in Fiji as 40.25% of the sample already make online purchases at least once a month. 37.80% of the respondents chose ease of purchase as a contributing factor in transitioning from traditional walk in shop purchasing to online e-commerce shopping with the second highest factor being accessibility with 17.07%.

9. Conclusion

It can therefore be concluded that before a certain type of technology is introduced, it is pragmatic to conduct market research to gauge the views of the receivers of the technology. This may also enable businesses to develop on features that matter rather than investing in technology which may not be widely accepted. The future of e-commerce to strive in Fiji seems beyond the horizon as many advances in technology will need to be made to gain the trust of consumers and decrease the perceived risks.

9.1 Recommendation

This study suggests that for Fijian businesses to venture into e-commerce, the platform must be secure to gain consumers' trust and reduce consumers' perceived risk of e-commerce. Together with this, providing sufficient information on e-commerce site and making sites easy to use will positively impact consumer's behavior of trusting and transacting online.

9.2 Further research opportunities

This research was conducted with a time constraint, thus making it difficult to obtain a larger sample. The study was conducted in a developing country – Fiji, thus generalizing the results in a rapidly growing world may be difficult. Further research may be conducted with a larger sample size to give a true depiction of the Fiji population. The hypothesis which was rejected was done so with a low correlation coefficient, hence further research can be conducted to verify and validate the results for this hypothesis.

Acknowledgements

For the purpose of this research we thank our class mates, work colleagues, friends and family who provided their perceptions through the online survey that greatly assisted the research, although they may not agree with all of the interpretations/conclusions of this paper.

We would also like to show our gratitude to Dr. Sam Goundar, Senior Lecturer – Information Systems, The University of the South Pacific for sharing his wisdom with us during the course of this research.

Author details

Sam Goundar[1*], Kunal Lal[1], Ashmita Chand[1] and Pranav Vyas[2]

1 The University of the South Pacific, Suva, Fiji

2 Smt. Chandaben Mohanbhai Patel Institute of Computer Applications, Charotar University of Science and Technology (CHARUSAT), Changa, Gujarat, India

*Address all correspondence to: sam.goundar@gmail.com

References

[1] Wang, P., Chaudhry, S., & Xu, L. D. Introduction: Advances in e-business engineering and management. Information Technology and Management, *17*(3), 199-201. 2016, September.

[2] Internet World Stats. *World Internet Users Statistics and 2018 World Population Stats*. (Miniwatts Marketing Group). 2017, December 31. Retrieved August 16, 2018, from Internet World Stats: https://www.internetworldstats.com/stats.htm

[3] Purcell, F., & Toland, J. (2004). Electronic commerce for the South Pacific: A review of e-readiness. electronic commerce research, *4*(3), 241-262.

[4] Purcell, F., Toland, J., & Huff, S. L. (2004). The potential of e-commerce for remotely located SMEs: Case studies from Samoa. In *Electronic commerce in small to medium-sized enterprises: Frameworks, issues and implications* (pp. 86–106). IGI Global.

[5] Molla, A., Taylor, R., & Licker, P. S. (2006). E-commerce diffusion in small island countries: The influence of institutions in Barbados. The Electronic Journal of Information Systems in Developing Countries, *28*(1), 1-15.

[6] Saini, A., & Johnson, J. L. (2005). Organizational capabilities in e-commerce: An empirical investigation of e-brokerage service providers. Journal of the Academy of Marketing Science, *33*(3), 360-375.

[7] Wu, J. H., & Hisa, T. L. (2008). Developing e-business dynamic capabilities: An analysis of e-commerce innovation from I-, M-, to U-commerce. Journal of Organizational Computing and Electronic Commerce, *18*(2), 95-111.

[8] Pavlou, P. A. (2003). Consumer acceptance of electronic commerce: Integrating trust and risk with the technology acceptance model. International journal of electronic commerce, *7*(3), 101-134.

[9] Gefen, D., Rao, V. S., & Tractinsky, N. (2003, January). The conceptualization of trust, risk and their relationship in electronic commerce: The need for clarifications. In *null* (p. 192b). IEEE.

[10] Wang, Y., & Lin, F. R. (2006, October). Trust and risk evaluation of transactions with different amounts in peer-to-peer e-commerce environments. In *2006 IEEE International Conference on e-Business Engineering (ICEBE'06)* (pp. 102–109). IEEE.

[11] Davis, F. D. (1989). Perceived usefulness, perceived ease of use, and user acceptance of information technology. MIS quarterly, 319-340.

[12] Klopping, I. M., & McKinney, E. (2004). Extending the technology acceptance model and the task-technology fit model to consumer e-commerce. Information Technology, Learning & Performance Journal, *22*(1).

[13] Lallmahamood, M. (1970). An examination of IndividualÃ¢ Â Â s perceived security and privacy of the internet in Malaysia and the influence of this on their intention to use E-commerce: Using an extension of the technology acceptance model. The Journal of Internet Banking and Commerce, *12*(3), 1-26.

[14] Johar, M. G. M., & Awalluddin, J. A. A. (2011). The role of technology acceptance model in explaining effect on e-commerce application system. International Journal of Managing Information Technology, *3*(3), 1-14.

[15] Wu, J. H., & Wang, S. C. (2005). What drives mobile commerce?: An

empirical evaluation of the revised technology acceptance model. Information & management, *42*(5), 719-729.

[16] Venkatesh, V., Morris, M. G., Davis, G. B., & Davis, F. D. (2003). User acceptance of information technology: Toward a unified view. MIS quarterly, 425-478.

[17] Ajzen, I. (1991). The theory of planned behavior. Organizational behavior and human decision processes, *50*(2), 179-211.

[18] Fishbein, M. (1979). A Theory of Reasoned Action: Some Applications and Implications.

[19] Fraumeni, B. M. (2001). E-commerce: Measurement and measurement issues. American Economic Review, *91*(2), 318-322.

[20] Chaudhuri, P., & Kumar, A. (2015). Role of Digitization and e-Commerce in Indian Economic Growth: An Employment Generation Perspective. *Research Gate. December*.

[21] Terzi, N. (2011). The impact of e-commerce on international trade and employment. Procedia-social and behavioral sciences, *24*, 745-753.

[22] Hasan, M., & Huda, M. N. (2013). E-commerce challenges, solutions and effectiveness perspective Bangladesh. International Journal of Computer Applications, *70*(9).

[23] Reserve Bank of Fiji. Financial services demand side survey Republic of Fiji. Suva. 2015.

[24] Wiedmann, K., Hennigs, N., Varelmann, D., & Reeh, M. Determinants of Consumers' Perceived Trust in IT-Ecosystems. 2010.

[25] Samadi, M., Yaghoob-Nejadi, A. A Survey of the Effect of Consumers' Perceived Risk on Purchase Intention in E-Shopping. 2009.

[26] Faqih, K. M.Exploring the Influence of Perceived Risk and Internet Self-efficacy on Consumer Online Shopping Intentions: Perspective of Technology

[27] Davis, F. Perceived Usefulness, Perceived Ease of Use and User Acceptance of Information Technology. Mis Quarterly 13(3), 1989, September.

[28] Hamid, Adnan., Razak, F., Bakar, A., Abdullah, W. The Effects of Perceived Usefulness and Perceived Ease of Use on Continuance Intention to Use E-Government. 2015.

[29] Fichter, K. E-commerce - sorting out the environmental Consequnces. Journal of Industrial Ecology, 6(2), 27. 2003. Retrieved from http://www.dartmouth.edu/~cushman/courses/IE-at-UNG/eCommerceConsequences.pdf [Accessed: 18-10-01]

[30] Abbass, S. M., & Ibrahim, O. B. A Model for developing online verification among E-commerce consumers. Journal of Theoretical and Applied Information Technology, 55(1), 39-52. 2013, September 10.

[31] Jones, K., & Leonard, L. N. Factors influencing Buyer's Trust in Consumer-to-Consumer E Commmerce. Journal of Computer Information Systems, 54(4), 71-79. 2014.

[32] Napompech, K. Factors driving consumers to purchase clothes through E-commerce in social networks. Journal of Applied Sciences, 14(17), 1936-1943. 2014.

[33] Muneer, A., Razzaq, S., & Farooq, Z. Data Privacy Issues and Possible Solutions in E-commerce. *Journal of Accounting & Marketing, 7*(3), 1-3. (n.d.). Retrieved 10 24, 2018, from https://www.omicsonline.org/open-access/data-privacy-issues-and-possible-

solutions-in-ecommerce-2168-9601-1000294.pdf

[34] Leitch, S., & Warren, M. Ethics and Electronic Commerce. 2002. Retrieved from http://crpit.com/confpapers/CRPITV1Leitch.pdf. [Accessed: 18-10-15]

[35] Nguyen, K. Business Ethics in E-commerce. 63. 2016. Retrieved from h ttps://www.theseus.fi/bitstream/hand le/10024/119487/Final%20Thesis-Kha nhNguyen.pdf?sequence=1. [Accessed: 18-10-20]

[36] Heijden, H. V., Verhagen, T., & Creemers, M. Understanding online purchase intentions: contributions from technology and trust perspectives. *European Journal of Information Systems, 12*, 41-48. 2003. Retrieved 10 23, 2018, from https://link.springer.com/content/pdf/10.1057/palgrave.ejis.3000445.pdf

[37] Butt, I., Tabassam, S., Chaudhry, N. G., & Nusair, K. (2016, August). Using technology acceptance model to study adoption of online shopping in an emerging economy. Journal of Internet Banking and Commerce, *21*(2), 2-18.

[38] ALraja, M. N., & Aref, M. Customer acceptance of E-commerce: Integrating perceived risk with TAM. International Journal of Applied Business and Economic Research, 13(2), 913-921. 2015.

[39] Broutsou, A., & Fitsilis, P. (2012). Online trust: The influence of perceived Company's reputation on consumers' trust and the effects of trust on intention for online transactions. Journal of Service Science and Management, 5, 365-372.

[40] Vasileiadis, A. Security concerns and Trust in the Adoption of M-commerce. Social Technologies, 4(1), 179-191. 2014. Retrieved from https://www.mruni.eu/upload/iblock/d76/ST-14-4-1-12.pdf. [Accessed: 18-10-22]

[41] Pokki, S. Web usability in e-commerce - usability evaluation of 4 web shops. Lahti University of Applied Sciences. 2016. Retrieved from https://www.theseus.fi/bitstream/handle/10024/119628/Pokki_Sini.pdf?seque nce=1 . [Accessed: 18-10-20]

[42] Renko, S., & Popović, D. Exploring the Consumers' Acceptance of Electronic Retailing using Technology Acceptance Model. 29-40. 2015, May 27.

[43] Venkatesh, V., & Davis, F. D. A theoretical extension of the technology acceptance model: Four longitudinal field studies. Management Science, 46 (2), 186-204. 2000, February.

[44] Vasileiadis, A. Security concerns and Trust in the Adoption of M-commerce. Social Technologies, 4(1), 179-191. 2014. Retrieved from https://www.mruni.eu/upload/iblock/d76/ST-14-4-1-12.pdf. [Accessed: 18-10-22]

[45] ALraja, M. N., & Aref, M. Customer acceptance of E-commerce of E-commerce: Integrating perceived risk with TAM. International Journal of Applied Business and Economic Research, 13(2), 913-921. 2015.

[46] Visser, P. S., Krosnick, J. A., & Lavrakas, P. J. Survey Research. (n.d.). Retrieved October 24, 2018, from https://web.stanford.edu/dept/communication/faculty/krosnick/Survey_Research.pdf

[47] McLeod, S. (Ed.). (2017). Qualitative vs Quantitative Research | Simply Psychology. Retrieved 12 12, 2018, from SimplyPsychology: https://www.simplypsychology.org/qualitative-quantitative.html

[48] Hair, J.F., Black, W.C., Babin, B.J., & Anderson, R.E., 2010, Multivariate Data Analysis: A Global Perspective, New Jersey, USA: Pearson Education Inc.

[49] Chin, W.W., 1998, The Partial Least Squares Approach to Structural Equation Modeling, on Modern Methods for Business Research, G. A. Marcoulides (editor), New Jersey, USA: Lawrance Erlbaum Associates.

[50] Lee, S., & Koubek, R. J. (2010). The effects of usability and web design attributes on user preference for e-commerce web sites. Computers in Industry, *61*(4), 329-341.

[51] Bertea, P. (2010). Scales for measuring perceived risk in E-commerce-testing influences on reliability. *Management & Marketing-Craiova*, (1s), 81-92.

[52] Sutton, S. G., Khazanchi, D., Hampton, C., & Arnold, V. (2008). Risk analysis in extended enterprise environments: Identification of critical risk factors in B2B e-commerce relationships. Journal of the Association for Information Systems, *9*(3–4), 151-174.

[53] Hidayanto, A. N., Saifulhaq, H., & Handayani, P. W. (2012, June). Do consumers really care on risks in online shopping? An analysis from Indonesian online consumers. In *2012 IEEE International Conference on Management of Innovation & Technology (ICMIT)* (pp. 331–336). IEEE.

[54] Faqih, K. M.Exploring the influence of perceived risk and internet self-efficacy on consumer online shopping intentions: Perspective of technology acceptance model. International Management Review, 9(1), 74. 2013.

[55] Gefen, D., Karahanna, E., & Straub, D. W. Trust and TAM in online shopping: An integrated model. MIS Quarterly, 27(1), 72. 2003, March.

[56] Bugembe, J. Perceived Usefulness, Perceived Ease of Use, Attitude and Actual Usage of a New Financial Management System: A Case Study of Uganda National Examinations Board. 2010.

[57] Pavlou, P. A. Consumer acceptance of electronic commerce: Integrating trust and risk with the technology acceptance model. International Journal of Electronic Commerce, 7(3), 101-134. 2003. Retrieved from https://doi.org/10.1080/10864415.2003.11044275. [Accessed: 18-10-24]

Chapter 3

Digitized Educational System

Suppanunta Romprasert and Jirayut Monjagapate

Abstract

Today's society, economy, and community are involved in terms of information technology usage. Thai university is a national university that has been highly supported by the Thai government. The term "autonomous university" refers to being independently organized under "bureaucratic frameworks." In other words, being an autonomous university lets the university to re-engineer itself to stay up-to-date for current market demands for labor. Most universities are already autonomous universities, but the system and structures still remain inefficient. Papers are used for records. Thai university has not kept records in soft copies. It is difficult to re-engineer university system because, to implement a new structure, it incurs huge costs. To digitize the university system, technologies, training program, and expertise are required. Moreover, online education and mobile students seem to be far though technology has been fundamentally implemented. Thai university mostly offers traditional classes, which it requires the students to attend with all paper works for the entire period. Active learning that has always been mentioned by government has still not been completely applied. According to these reasons the authors would like to recommend digital instruments, a way to re-engineer university systems, and other factors to support better educational systems.

Keywords: digitization in public, management, organization, formal institution, digital governance

1. Introduction

Internet of Things (ToT) is a form of Internet connecting into physical devices and objects. Internet of Things is a combination of several technologies such as real-time analysis and machine learning. Everyone from everywhere in this world can connect to each other via the Internet. Interactions are mostly done from the World Wide Web. It is interesting that nowadays, the Internet can not only connect to the web but can also connect to smart card, security, online banking, health care, education, and so on.

In addition, the Internet of Things can generate chances for new services and sales, which can be more efficient for the business sector. For the government, according to economic growth factors, there are five main pillars for policy making: (1) hard infrastructure; (2) soft infrastructure; (3) service infrastructure; (4) digital economy; and (5) digital society and knowledge. The digital economy seems to be outstanding these days because the economic growth no longer depends on production and consumption or touchable goods; rather, it additionally depends on intangible goods.

The educational system in current economic and social conditions are forced to use as in the Age of "Information Technology". The growth rate is increasing in the area of knowledge because of the trend "lifelong learning," which the old style may not be able to support the creation of a career for a new generation anymore. Professionals must always adjust and be ready to learn. The industry must develop the skills of employees to be appropriate.

The learning style that is suitable for the information age is a model that can bring technology to use by learning appropriately with the interests and abilities of each individual. This resulting is called "New Learning Style (NLS)" such as massive open online courses (MOOCs) for the masses or other learning channels whether they are videos, audios, social media, or games. It shows a good opportunity for the new generations to choose a learning style that suits them the most.

Will this method be widely used in Thailand consistent with the behavior of Thai's new generation? There is a high chance that the new generation will be able to access it, including causing learning changes in the near future.

2. Digitized educational system

The knowledge that young generations need more is "working skills and life skills" in daily life, respectively, by means of additional knowledge using media such as videos, audios, social media, etc. with Internet connectivity. For the comments from young generations on using online media, it is recommended to have a variety of learning topics especially for the classroom in particular applied in various fields of subjects for instance, in Economics of Information Technology; it is a study of overall economics in the age of technology, structure of markets and information technologies, product and price differentiation method, costs associated with changing technologies, economies of scales, effects of network, product standards, the effects of linked product system, effects of development an transformation of technology toward the economic and industrial policies. In Creative Economy, students learn ideas of economic motivation on the fundamental of integrating between education, creativity and the use of intellectual property to link with culture and accumulated knowledge of sociology, technology and new inventions. In International Business Management, it is a study of basics theory of international business management, international merger and acquisition, opportunities for export and import, analyzing strategies and competitiveness of the international organizations, etc.

Digitized educational system can be used to present as creative media providing many pictures, but less content. To support on this issue, instructors should have knowledge of the subjects taught using an online media to make it more interesting and more accessible for students.

However, the major problems and obstacles in learning are related to attitude differences, individual tastes, and judgment of teachers and learners. These obstacles are a result of development of analytical thinking skills to promote learning via a combination of media-interactive learning such as social media videos, together with traditional media such as movies. Inspiration and diverse learning are main factors that can actually be a motivation to learn for the younger generation.

2.1 Digitized teaching and learning

When implementing each learning platform, using the media selection suitable for the needs and context of the end user should be considered. One approach outstanding today is on preparing to benefit from the learning materials fully with examples of ideas to know "Massive Open Online Course (MOOCs)" [1]. To study MOOCs, learners should be provided with a mentor system, and MOOCs with industry sector practice must be implemented. With technological advances, coupled with economic and social dynamics, causing changes in all learning channels, technology is known as the key driver, whether it is artificial intelligence (AI) that causes significant changes in every platform or otherwise. Automatic chat program will be the other channel, which can be divided into the following main groups: gathering knowledge and exchanging knowledge between students with diverse knowledge and experiences.

Some part of the world, using the technology of games like AR VR to promote learning and in the same time using social media for the benefit of learning like technology and tools to help learning activities. Moreover, promoting the use of audio media creates equality in learning for all groups of people. It should encourage Thai people to know the source of Thai audio books for benefits.

2.2 Project writing and digital data evaluation

Not only the university but also every government institution, all public organizations, and private sectors must follow bureaucratic frameworks to do the paper work and follow the bureaucratic forms. When a project is coming out, every procedure of the project must be written and recorded as paper work. The importance of paper work is to serve as evidence and all papers should be coded and recorded in the book. This is to confirm that every step is seen by every department and checked many times.

For example, for a project proposal, first of all the budget of that activity will be planned before the year of budget (for public institutions that follow the bureaucratic frameworks, the year of budget will begin on October 1st, this year, and will end on September 30th, next year). The plan's details need to be discussed at the conference, so papers will be printed to distribute to every member at the conference. Next, when it is time to start the project, the project needs to draw on many topics such as declaration of intentions, costs, evaluation, and schedules. After the project has been approved, all the papers are collected and kept as evidence. When the project ends, an evaluation is required. The evaluation can be in a paper form or a writing form. The number of papers in this process depends on the number of participants. All processes that have been stated are made by papers.

Figure 1 illustrates the process of a project procedure; every step requires paper work. After the project ends, all papers are kept as physical evidence. No or few papers are scanned and transferred into a digital file. After that, all papers become

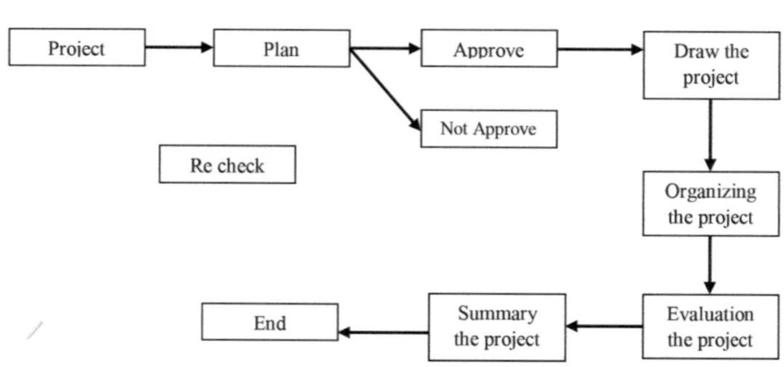

Figure 1.
Brief process of project procedure.

physical "dark data" or the information that we no longer use, but we cannot eliminate them completely because they will, sometimes, be used later. However, the amount of papers causes the problem of managing. Also, it becomes difficult to search the old papers. The importance of paperwork: In the bureaucratic system, it is important to keep papers because they can be used as evidence and references for other departments, private institutions, and citizens. The papers also can be used for legal protection. In addition to private sectors, some companies do not have sufficient space to keep all the papers, so they quickly adapt to the new technologies to change the forms from physical files to digital files.

However, there are problems of having paperwork. Because these papers can be used by many people in the organization, the papers can be lost or moved. Possibly, in terms of human capital, people lack the knowledge of managing papers. In the worst case, when the organization loses papers, it is hard to find them or recreate paperwork, and cost for recovery is incurred. Hence, it is important to transfer papers into digital forms to facilitate the organization. In governmental institutions, paperwork still exits, and some of the governmental institutions fully have paperwork or few digital files. It is a risk to have only paperwork.

The Faculty of Science, Srinakharinwirot University, Thailand, [2] announced that one faculty consumes 1700 ream or 850,000 pages annually. It costs about almost 200,000 baht per year Or 6286 US dollars. If the amount of paper consumption reduces, the environment will be saved and the cost of operation inside the department can be minimized. **Figures 5** and **6** show the change in society in terms of technological approach and many organizations still have remained the same. To show the evidence, according to [3], overproduction of paper cause significantly to pollution, deforestation, and greenhouse gas emissions.

For the recommendation, we put the data of the projects and other information into the digital system, it will consequently reduce the cost of operation because data are transferred from physical to digital. Moreover, it makes channels of access of data better and easier, and it can reduce cost of transportation because data can be accessed everywhere. This would be beneficial to all government, business cooperation, and people in both macroeconomic and microeconomic perspectives [4]. **Figure 2** shows a cloud for education. Users can log in to their accounts and upload documents, and other users who are authorized can also download the information. "*Cloud is the practice of using a network of remote servers hosted on the Internet to store, manage, and process data, rather than a local server or a personal computer.*"

Moreover, if the data are transferred into the digital forms, it can be copied unlimitedly when we want to distribute. This can lead to *zero marginal cost*. The zero marginal cost refers to marginal cost of producing one additional unit that does not make any additional cost. The cost that we need to handle is the cost of the system [5].

2.3 Merged mail and digital worked assignment

The importance of the educational sector must be those who play an important role in learning management for "Quality of Learners (OL)". OL is to enable students to manage their own learning. There are push-factors to positively change

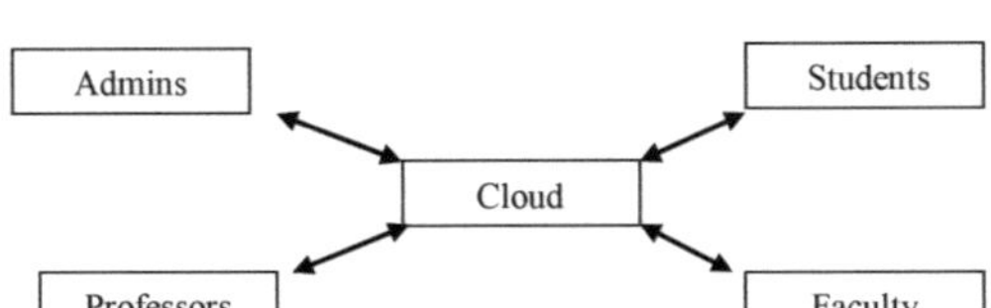

Figure 2.
Cloud for education.

the students. It may come from technological development, social changes, career changes. New generation children must have the ability to work [6, 7]—if there is a need for both advanced skills and critical thinking to support the future world.

For the proper educational system, students should be assigned homework properly. The assignment needs to be adapted to meet the needs and necessities of Thai society and the young generation, focusing on developing learners to be able to create innovation and create work together with personal skills [8]. Linking the order process with innovation is essential in every part of the educational system and that must be immediately and quickly changed together with the assignment through exchange of learning.

Developing a basic service platform of communication and providing the application can make more convenience and efficiency to all people in the system as follows: E-School or Electronic-School: Faculty of Technology used to improve public administration efficiency combining with increase transparency and reliability; and Smart School.

E-school has been developed to the point that at least two groups such as to support staff and to enable students to connect to work learning and teaching without adhering to the boundaries of duties and responsibilities in accordance with the mission of the faculty, but taking into account the benefits of the service provider as the location [9].

The goal is to deliver quality services to students in the form of "digital interaction" between departments such as Student2University (S2U); University2University (U2U); University2Student (U2S), and Student2Student (S2S).

An example for convenience is that students are allowed to complete their assignment anywhere as long as it is done before the due date. Students do not have to hand in their assignment in the professor's room. The file that they submit is an electronic file. This does not waste a natural resource like paper. In Section 2.2, the effects of using a large number of papers that could cause natural damages has already been demonstrated.

Figure 3 shows details of an assignment via Google Classroom (free), date of submission, and date of post by professor. This platform not only helps professors to check the date of online submission from students and but also helps to decrease paper work and wasted papers. Also, all data are recorded. Interestingly, the file that students submit can be checked and reviewed by other organizations such as https://www.turnitin.com/. The website can be a helpful assistant for professors for checking plagiarism. The website can check the information that has pdf, doc, and other electronic files. On the contrary, papers are difficult to recover when the paper is damaged or lose, and it is also difficult to adapt with other program or website.

In an academic field, plagiarism is considered as a serious crime. Almost all institutions set serious regulations against plagiarism. **Figure 4** demonstrates the similarity of the student's paper, as shown by the Turnitin website. In general academic rule, similarity must not exceed 20–30% for undergraduate students, 10–20% for graduate students, and 10% for PhD students. This rule depends on the faculty and the professors' consideration. It is, therefore, noticeable that merging new technologies helps teachers and learners in many aspects of academic files.

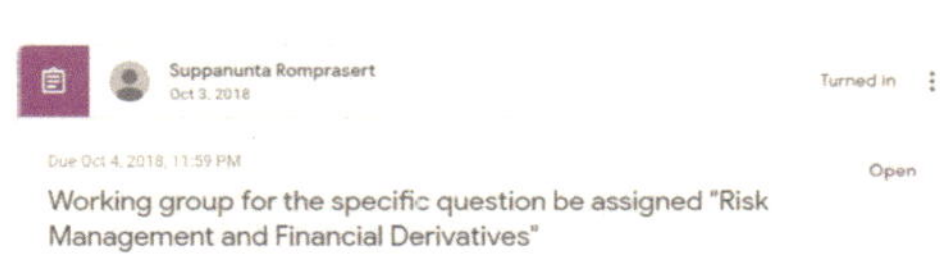

Figure 3.
Online classroom & assignment via Google Classroom.

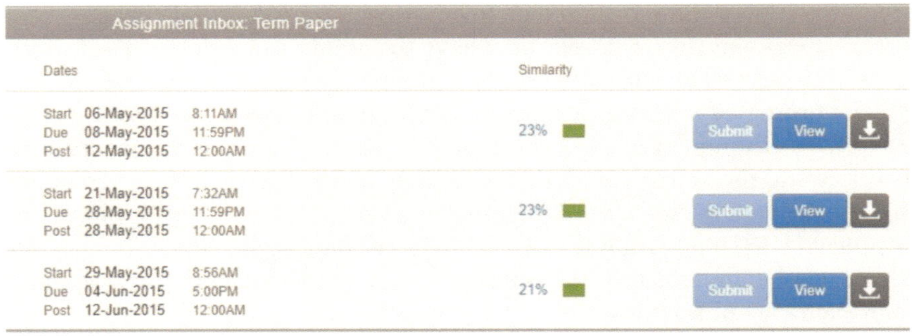

Figure 4.
Table of assignment submission via Turnitin.

This does not benefit the education system only. The example intentionally aims to show how systematic the task allocation is. The platform of digital task allocation can be applied to all governmental institutions and private sectors to increase work efficiency and productivity. There is interesting statistical evidence why we should adopt a digital workplace. According to [10], approximately 64% of employees accept a lower wage if they are allowed to work away from office. Online social network generates more than 7% productivity to organizations. When the organizations use the social media tools, it increases 20% of employee satisfaction.

2.4 Digital combining for teaching systems

Many institutions adopt computer games as one channel of teaching; AR and VR technologies in games are used to promote learning. Teachers try to use social media for the benefit of learning unlike in the past and also use audio media creating equality in learning for all groups of students. Future classrooms should be brought up to speed to change educational system [11]. There are three significant aspects: The first is children; the environment around the children involves a lot of technology, which has both disadvantages and advantages; however, most disadvantages are causing children to become more hyperactive and autistic. The second is the use of smartphones all the time, adversely affecting their own health. The third aspect is the positive development of gadgets including the rapid emergence of features resulting in the behavior of people changing and making the classroom more attractive.

In recent times, the young generation thinks that having a classroom or learning from school is unnecessary. The young generation can learn via smartphone causing the question 'If we allow technology to play a role and learn too much classes or platforms, knowledge based on schools or universities, will be depleted and important. Earlier there would be awareness about the educational reform system, but focused on teacher development. No one could talk about the classroom, where the teachers were irreplaceable.

The teacher is like a warrior fighting to win, but he or she must make a good army commander, but we forgot to develop weapons for teachers—they are left with a weapon that is like the rusting spear to the warrior. When he or she does a career related to the design experience, there is a lot of knowledge about using modern technology media. The right discussion will meet good results. The variables normally use to comment be skills, experiences and knowledge which those should be used to make the classroom considered as "a weapon development for teachers" in the modern age.

In addition to this classroom set, there will be special characteristics that allow teachers and students to learn together. The technology available in the room will help spark creativity that can be lead to displaying creative ideas [12]. For example, the room has a laser projector projected in bright light with an image extension technology. Can those be connected to a creative environment for students to see the real thing? The actual size is better than sitting on the chair and reading only the textbook. To support the technology that will plug and support dimensionless classroom activities, teachers and students must hold hands and walk together. There are so many activities that the teachers can change the classroom and make it look like an exhibition, allowing students to work on their own experiences. The process of real work will help students to discover their identity and aptitude including the practicing skills to deal with tasks.

2.5 Teaching and learning with digital systems

High school and university are different. University students focus deeper in their interested fields than when they were high school. University's structures: Instructors or professors are responsible for the students in many ways. It is more than just coming to teach and check everything the students finish. The professors have to prepare lecture notes, research, do quality assurance of the course, draft and organize the university activities, be an advisor, and teach. That is why professors do not just only teach and do research as many people believe they should do.

Figure 5, [13], shows the percentage of population using the Internet for the period 2010–2016. In 2010, only 22% of Thai population accessed the Internet. Six years later, in 2016, the number of Thai population accessing the Internet is 47%, which is higher than the 45% of the world population. [14] **Figure 6** also goes in the same direction as **Figure 5** and shows the number of mobile cellular subscriptions

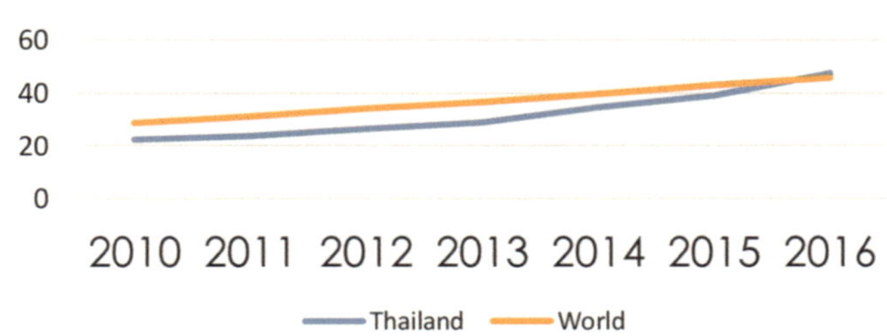

Figure 5.
Individuals using the Internet (% of population).

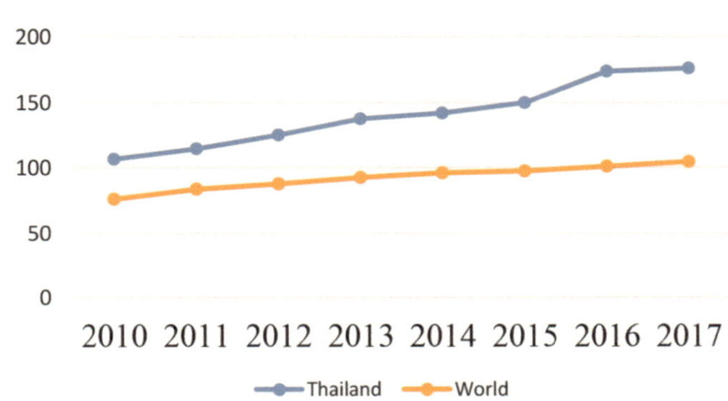

Figure 6.
Mobile cellular subscriptions (per 100 people).

that has been increasing. Especially, in Thailand, the number has long been raising beyond the world's average. In other words, 1 Thai person probably has more than 1 mobile phone.

It is interesting that Thailand has adapted to the Internet and smart phones faster than the rest of the world's average. From this statistical evidence, we can expect that Thailand would have probably transferred analogue system to the digital system. However, for the educational system, Thailand has just started to turn the education system into the digital system.

To illustrate the statement above, almost every Thai university is a national university, which has been highly supported by Thai government to become "*autonomous university.*" The term "autonomous university" refers to be an independent organization under the bureaucratic framework. In other words, being an autonomous university lets the university to re-engineer itself to be stay with the flow and up-to-date based on the current market demands, which are changing all the time and faster than before. However, online education and "mobile students" seem to be far from here though it should have been fundamentally implemented already. Thai universities have long offered traditional classes that students have to attend, and some courses still use paper works for the entire period of the course. Because of being autonomous universities, some universities have made their system stricter than before. The term "strict" means that every procedure and process inside the university must be recorded and printed out. Some procedures that should not be with the paper work still appear. The paper work actually takes time and effort. The paper work can cost a high opportunity cost relative to other work that the Thai professor should do. The opportunity cost means the loss of gain from other alternatives. For example, you spend 4 hours for the paper work, but you can exercise and read some good books for 4 hours.

Moreover, professors have various tasks in one day. As it has been stated before, Thai professors do not just teach and do research. In addition, every student must have a professor as an advisor for his entire student life. Professors must spare their time for other students who enroll to their course.

Figure 7 illustrates from the survey that university students make an appointment and request on meeting with professors face-to-face. The results from survey are interesting that every Thai university student chose face-to-face. However, for other ways to communicate with professors, we have new technologies to make easier communication such as Line Application and E-mail, but less than or equal half of the numbers of students use these applications.

When we compare the growth number of accessibility of Internet and mobiles with the way students contact their professors, the number of digital tools and digital services that students use for contacting their professors go in different directions. From previous illustrations, we can see some inefficiencies from not adapting new technologies. For example, in students' side, students do not much adopt the digital technology in the educational way, and it can convey to how

The Ways Thai University Students Contact their Professors							
Gender		Face-to-Face	Line Application	Email	Facebook	Personal Telephone	University Website
Male	Number	10	2	4	2	0	1
	Percentage	100%	20%	40%	20%	0%	10%
Female	Number	16	8	2	1	2	1
	Percentage	100%	50%	12.5%	6.25%	12.5%	6.35%
Total	Number	26	10	6	3	2	2
	Percentage	100%	38.5%	23.1%	11.5%	7.7%	7.7%

Figure 7.
The ways Thai university students contact their professors.

professional they are. For professors' point of view, it is difficult to manage their schedule because students come to ask for face-to-face appointments. Sometimes, professors can possibly get distracted when they are working in their room. Additionally, this can be a major problem for professors if the class they teach contains a large number of students.

This, therefore, comes to the recommendation that the university should promote and announce digital techniques to the class and add some regulations. This can reduce the difficulty of professors. Students also benefit from this such as knowing how to formally contact with professors, becoming more professional, and using a way to communicate while they are in the university to apply for the jobs. Some research points out that using digital goods and digital services assists better learning experience as the digital goods and services represent a medium for communication [15].

However, the table of survey can be conveyed beyond the university concerns. Educational institutions should fundamentally be a pusher to everyone to perceive and learn to use new technologies. Also, they should provide some skills to all learners. The learners are not limited to students only.

3. Conclusion

Thai education guidelines should be managed to solve problems appropriately and creatively. It will be able to proceed in many ways, especially integrating the content of digital learning in the context. Knowledge management in institutions has a variety of characteristics, depending on the context of the organization. Some institutions look to manage their knowledge when they are related to strategies or practices, so they focus on knowledge generation and knowledge storage. If it focuses on practice, we can apply the knowledge by integrating the factors involved in information technology conjunction with the importance of personnel in the institutions. Therefore, it is a form of knowledge management from the foundations by bringing ideas from different sources to adapt in the context of their own institutions.

Acknowledgements

I am grateful to all those I have had the pleasure to work with during writing of this book chapter. Each of the members of my teaching assistant team, especially Mr. Jirayut, has provided me extensive professional guidance and supported me a great deal with scientific academic articles in general. I would especially like to thank MS. Nina Kalinic Babic, Author Service Manager. As a supporter, she has given us credit for writing the book chapter.

The members of my family have been more important to me in the pursuit of this book chapter. I would like to thank my parents whose love and guidance are with me in whatever I pursue. They are the ultimate role models. Most importantly, I wish to thank my loving and supportive husband, Sa-ard, and my single wonderful child, Thanapat, who provide unending inspiration.

Author details

Suppanunta Romprasert* and Jirayut Monjagapate
Srinakharinwirot University, Bangkok, Thailand

*Address all correspondence to: suppanunta@g.swu.ac.th

References

[1] Kaplan AM, Haenlein M. Higher education and the digital revolution: About MOOCs, SPOCs, social media, and the cookie monster. Business Horizons. 2016;**59**:441-450. DOI: 10.1016/j.bushor.2016.03.008

[2] Faculty of Science. Srinakharinwirot University. 2016. Paperless Manual

[3] Mukete B, Yujun S, Zama E, Monono S. Paper consumption and environmental impact in an emerging economy. Journal of Energy, Environmental & Chemical Engineering. 2016;**1**(1):13-18. DOI: 10.11648/j.jeece.20160101.12

[4] Yadav K. Role of cloud computing in education. International Journal of Innovative Research in Computer and Communication Engineering. 2014; **2**(2):3108-3112. Retrieved from https://www.researchgate.net/profile/Nayan_Kadam/post/What_do_you_think_about_the_use_of_new_online_information_technologies_in_the_education_process/attachment/5c57ae113843b0544e63e4eb/AS:722421288017920@1549250065344/download/cloud_education.pdf

[5] Øverby H, Audestad JA. Digital Economics how Information and Communication Technology Is Shaping Markers, Business, and Innovtion. 1st ed. Scotts Valley, California: CreateSpace Independent Publishing Platform; 2018

[6] Coleman L. The role of school administrator as knowledge manager: A process for school improvement [Dissertation]. Illinois, US: Northern Illinois University; 2008

[7] Collison C, Parcell G. Learning to Fly. UK: Capstone; 2004

[8] DeTienne K, Dyer G, Hoopes C, Harris S. Toward a model of effective knowledge management and directions for future research: Culture, leadership and CKOs. Journal of Leadership and Organizational Studies. 2004;**10**(4): 26-43

[9] Dixon N. Common Knowledge, how Companies Thrive by Sharing What they Know. Boston: Harvard Business School Press; 2000

[10] The Digital Workplace: Think, Share, Do Transform Your Employee Experience. Deloitte Touche Tohmatsu Limited ("DTTL"). https://www2.deloitte.com/content/dam/Deloitte/mx/Documents/human-capital/The_digital_workplace.pdf

[11] Frappaolo C. Knowledge Mangement. New Jersey: John Wiley and Sons; 2006

[12] Marquardt M. Building the Learning Organization. California: Davies Black; 2002

[13] Individuals using the Internet (% of population). The World Bank 2019. https://data.worldbank.org/indicator/IT.NET.USER.ZS.

[14] Mobile cellular subscriptions (per 100 people). The World Bank 2019. https://data.worldbank.org/indicator/IT.CEL.SETS.P2

[15] Hassini E. Student–instructor communication: The role of email. Computers & Education. 2006;**47**: 29-40. DOI: 10.1016/j.compedu.2004.08.014

Chapter 4

Educational Technology: Relevance to a Fijian Classroom

Reginald Gani, Sharika Devi, Sam Goundar, Emmenual Reddy and Fatemeh Saber

Abstract

Educational technology has emerged as a necessary tool for classrooms in the developing world. Educational technology is the considered implementation of appropriate tools, techniques, or processes that facilitate the application of senses, memory, and cognition to enhance teaching practices and improve learning outcomes. It is defined as "the study and ethical practice of facilitating learning and improving performance by creating, using, and managing appropriate technological processes and resources." This research intends to identify educational technologies that are popular around the world and elaborate upon them. The research project further used a case study to sample schools in Fiji to determine their use of educational technologies in the Fijian classroom. Various uses, benefits, implementations, and challenges were identified together with the limitations of how educational technologies' relevance to the Fijian classroom could be affected. The findings indicate that the Fijian classroom is not quite ready to completely embrace educational technologies into their operations given the challenges are few but immense. Questions suggesting future activities were also identified.

Keywords: educational technology, e-Learning, teaching practices, Fijian classroom

1. Introduction

Education is the heart of any nation, and as the nation progresses, so must its educational delivery systems. For most part of Fiji's education system, the traditional approach has been utilized to implement a national curriculum. Looking at the rapid changes to the Fijian lifestyle in the recent years, especially in terms of technological infrastructure, communication, and the mobile revolution that has taken place, there remains a need to re-look at how this implementation and delivery of the Fijian education is taking place.

Educational technology is the considered implementation of appropriate tools, techniques, or processes that facilitate the application of senses, memory, and cognition to enhance teaching practices and improve learning outcomes [1]. Januszewski and Molenda [2] define education technology as "the study and ethical practice of facilitating learning and improving performance by creating, using, and managing appropriate technological processes and resources." Thus educational technology includes multifaceted implementations that are diverse in nature. These technologies do not only consist of computers and projection devices for teachers, but they include computers and portable devices for the learners, the hardware and software support systems to host and deliver learning resources, the technical expertise required to carry out implementation, and continued support and maintenance and the financial effort that would accompany all these activities.

The Fijian education system is a slow to change enterprise. Legacy practices remain the greatest challenge for any reform to handle. Teacher education, student comprehension, resource availability, and student access are all challenges to the implementation of any educational technology systems in schools. A key challenge that must be kept in mind is the "attitude towards technology and perceived usefulness of technology in teaching, the institutional cultural environment, as well as resources available to support uptake" [3].

Therefore, the purpose of this paper is to research on the educational technologies used in classrooms around the world and determine the availability of educational technologies to a Fijian classroom and the relevance of the available technologies in a Fijian classroom. Research by Ghavifekr and Rosdy [4] states that the integration of information, communication, and technology (ICT) will assist teachers to the global requirement to replace traditional teaching methods with a technology-based teaching and learning tools and facilities. Some of these tools could be made available to Fijian teachers to implement the syllabi that they are expected to deliver. Not every single technology that is availed to educators is feasible or logical given the socio-economic background of any educational institution or its students, but their relevance could be regulated to identify which ones are applicable to Fijian schools.

1.1 Purpose of study

The purpose of this research will be to ascertain if the Fijian classroom is ready to incorporate e-Learning concepts completely or even partially into its teaching and learning methods. The intention of this research is to outline the myriad of technologies and devices heralded for e-Learning in classrooms around the world and relate them to the Fijian classroom situation.

1.2 Research problem

This study would help understand where the Fijian classroom stands in comparison with classrooms around the world today in the case of the use of educational technologies and to determine their relevance through user feedback on acceptance of change.

1.2.1 Subproblems

The subproblems are as follows:

1. What are key educational technologies used in classrooms around the world today?

2. Are the educational technologies that are used around the world today available for the Fijian classroom?

3. How many of the key educational technologies available for the Fijian classroom are relevant for them?

2. Literature review

Educational technologies remain one of the major contributing factors to improve the performance of students as well as facilitate teacher in teaching and learning. Educational technology is not restricted to individual computer use. It can involve other equipment such as tablets and mobile devices as well as tools made available via the Web, such as the cloud and its applications [5, 6]. It is a known fact that the application of educational technology enhances skills and cognitive characteristics [7], and if such a tool exists, then it must be incorporated into the educational pedagogy.

While there are a number of educational technologies used in classrooms around the world, their availability and relevance of the available technologies in a Fijian classroom need to be looked at. This research is needed because as mentioned in earlier "Modernising an education system should work with both, the syllabi as well as the methods employed by teachers to implement them," the use of educational technologies plays a major role in the delivery of these syllabi that have been entrusted by the Ministry of Education [8, 9]. "Learning, teaching, and assessment enabled by technology require a robust infrastructure" [10]. This robust infrastructure is a reference to the educational technologies that can exist in any classroom.

Technology gives teachers the ability to tailor instructional materials and assessments to directly address their students' learning needs and offers access to more authentic material to assist in the development and delivery of lessons; they can potentially contribute to reshaping teaching and learning practices, and they might be able to replace a wide array of previous pedagogical tools, such as blackboards, textbooks, student and teacher notebooks, and laboratories [11, 12]. It is time that the "education system should soon give up on its obsession with textbooks and come out from the old-fashioned way of teaching students for an outcome-based education" [13].

This reality is visible all around the nation, where textbooks govern the delivery of education to students. With time, age, and budget that foot us with the rest of the world given our size, Fiji needs to start embracing innovative methods of delivery to prepare critical thinkers once students enter university, instead of beginners finding their feet with educational technology while they are undergraduates. An undisputable fact is that today's generation will grow up with a myriad of educational technologies around them and as such need to "cope with these changes and bridge the gap between the world outside school and inside school" [14]. The truth remains that "it is becoming increasingly easy for learners to bypass traditional bricks and mortar language schools and courses; in this scenario, the role of the teacher becomes that of guide, facilitator and consultant..." [15]. Thus, teachers need to be technologically prepared to guide students in the right direction and learn how to discover the correct information in the context of their delivery [16].

Lund et al. [17] have brought about a concern in teacher education with regard to helping students develop a profession-based digital competence relevant to teaching. While we herald the importance of so many benefits and needs for educational technology, an important fact that must be borne in our minds is the need to have teachers versed in using educational technology to prepare and implement lesson [16]. The entire exercise and investment would prove futile should the educator fail to comprehend the tools at their disposal.

Not only should we be concerned with implementing and using educational technology and claiming it as a yardstick for success, it is imperative that "educators need to learn how student learning changes with e-Learning, and how to alter their teaching methodologies with pedagogical approaches that take advantage of the opportunities afforded by-learning" [18–20]. Teaching using educational technology is not a one-way street, it is by far more complex than the traditional learning system given that the electronic methods appear as the third leg of the educational process and thus be balanced.

Finally, "in a high-tech economy, technological innovation needs to move hand in hand with investing in teachers, which is a far better long-term solution. Students will be made future ready through a blend of technology and teachers, the best of both worlds which will be needed to teach future generations" writes [16, 21].

2.1 Educational technologies' literature

See **Table 1**.

	Article ID	Topic area	Research approach	Data analysis	Major findings
1.	[6]	Educational technologies	Longitudinal data and student experience and expectation of technology (SEET) survey	Survey focuses on the current use and future use as well	Discovers that there was an increase in technologies for learning over 3 years
2.	[7]	Importance of educational technologies	Empirical research on educational technologies in classrooms	Importance of educational technologies in classrooms	The presence of technologies is growing in the classroom
3.	[10]	Changes in educational technology	Time series data analysis of national feedback to the Office of Educational Technology	Changes in the use of educational technologies in classrooms compared to 2016	Increase in the use of educational technologies in classrooms
4.	[12]	Educational technologies to deliver textbooks	Qualitative study	Delivery of texts using educational technologies	Increased access to textbooks for education
5.	[13]	Educational technologies are transforming education	Time series analysis of technological improvements in reaching an audience	Focus on the increased reach and content available for delivery using educational technologies	Astronomical increase in reach and content creation for curricula over the past years
6.	[14]	Educational technologies to innovate teaching	Review of journal articles	Compare innovative techniques	Innovative techniques are readily available to improve teaching and learning
7.	[15]	Current and future trends in technology	Review and study of educational technology trends over time	Comparison of past, present, and future technologies	Exponential increase in types and availability of educational technologies
8.	[17]	Professional digital competence	Research of educational technology reports	Compare reports on the need for digital competence	Greater need for digital competence for educators over the past years

	Article ID	Topic area	Research approach	Data analysis	Major findings
9.	[18]	Resurgence of e-Learning	Study on educational technologies over time	Comparison of reports over time to evidence the steady presence of educational technologies over the years	Despite expert predictions of the failure of educational technologies in education, it remains
10.	[21]	The future of education is ease of access and ease of learning using educational technology	Review of journal article	Technology is improving over the years to make delivery of education and access easier	Educational learning is slowly becoming the key delivery tool

Table 1.
Review of existing literature on educational technologies.

3. Educational technologies

Educational technologies have three domains of use [7]:

- Technology as a tutor (computer gives instructions and guides the user)

This involves the use of any form of technology to assist and educate a user through lesson/ lessons. This method is utilized to deliver self-paced education and is ideal for tertiary level courses and distance learning or to deliver syllabi to remote areas without any access to educator. This approach uses a drill and practice implementation whereby the user can keep repeating the lesson until the required mastery is achieved. An example of this would be a typing tuition application, allowing for self-paced, timed lessons. There are, of course, numerous advanced applications that are available for self-paced learning [22].

- Technology as a teaching tool

This educational technology approach is by far the more popular approach used around Fiji. It involves educators utilizing technology to deliver curriculum to learners through, for example, the use of Microsoft PowerPoint to present slides and portable document formats to supply notes, worksheets, activities, and textbooks. This requires deep engagement by the educator to prepare lessons for delivery. This, of course, could be substituted with ready designed courses that are provided for a fee [22]. This tool can be expanded into numerous hardware aids that exist for educators to utilize in classrooms. Some common examples of lesson delivery hardware equipment are interactive whiteboards, overhead data projectors with portable computers, and smart screens [23].

- Technology as a learning tool

These normally consist of computer hardware that is used by students to gain access to the resources that are prepared and shared by educators. These resources may be placed on a server, and the student is required to be part of the domain via proper credentials. Some examples of these tools are personal computers (PCs), laptops, tablets, and smartphones. These can also be used to gain access to tutorial activities for learners to access and progress academically [23].

3.1 Common types of classroom technologies

See **Table 2**.

	Technology	Use
1.	Electronic whiteboard	These are best used for group presentations. The board may be connected to a PC or a laptop
2.	Flipped learning	Students can use their devices to watch videos of lessons and then have discussions later
3.	Desktops and laptops	These more common devices can be used to gain access to the resources available for educational learning
4.	Projectors	Overhead data projectors allow an educator to share the contents of their screen to the class
5.	Video conferencing classroom technologies	Students in different locations can be present in classroom lessons in real time using technologies like Skype, Google Hangouts, and Zoom
6.	Mobile learning	Mobile devices are portable and allow access to classrooms and apps wherever/whenever needed. This medium is popular in distance learning
7.	Television	These can be used to play educational videos or demonstrations to share lessons
8.	Computer networking	Allow students to share resources and for teacher to monitor student activity on the shared network
9.	Distance learning	Students can gain remote access to classes if restricted by some event
10.	Virtual field trips	Students can go onto seemingly real-life environments, for example, Google classroom
11.	Word processing applications	Students' first exposure to technology in Fiji is Microsoft Word
12.	3D printing	Both students and teachers can design 3D object for everyday marking

Table 2.
Modern educational technologies [24].

4. Theoretical framework

The framework was drawn up in relation to evaluation of the adoption and use of educational technology in the Fijian classroom. The research was based on e-Learning in Fijian schools in terms of the following: implementation, benefits, challenges, and use (e-Learning for teaching or a complete implementation of e-Learning including both delivery and students' ends) (**Figure 1**).

The intention of this framework was to take a snapshot of the presence of e-Learning in some Fijian school. This was undertaken using a convenience sample method to identify the current state of e-Learning in the sample of schools. The implementation and challenges will be evaluated once the convenience sample was received. From the sample, the benefits and use would be identified as they would be different for each sample. While the generic ideas on the implementation, challenges, benefits, and use are easily available, the research pinpointed these to the frame of a Fijian classroom. The key outcomes of this research were to identify the

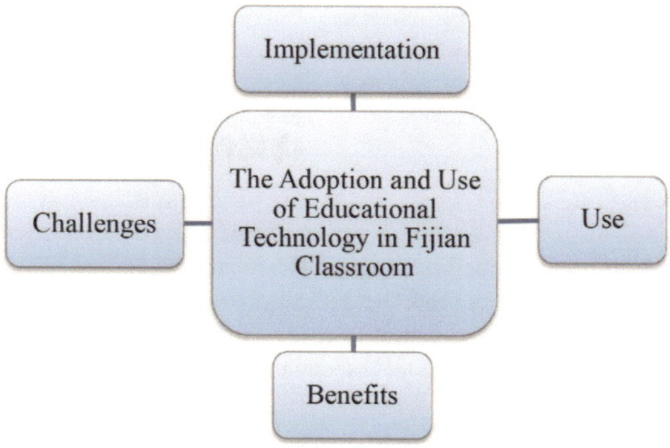

Figure 1.
Model.

depth of the adoption or e-Learning in the sample, benefits and challenges, and the current use of e-Learning.

5. Methodology

The approach to this paper was to review journal articles to discover and elaborate on educational technologies, its presence, and use. The types of educational technological hardware were also researched on and the more common ones identified. Following this, a type of nonprobability sampling method was used to gather data on the stance of some Fijian school on the use, implementation, benefits, and implementation for teaching and learning purposes. This method is referred to as convenience sampling. Convenience sampling (grab sampling, accidental sampling, opportunity sampling, haphazard sampling) is a type of nonprobability sampling in which people are sampled simply because they are "convenient" for researchers [25]. The survey questionnaire was delivered to respondents using an electronic questionnaire via www.surveymonkey.com. The sample was biased to schools in urban and suburban areas. This survey method is best used if a researcher desires to launch a pilot study for a case where minimal historical data or research is present. The responses of the survey were analysed under the theoretical framework based on implementation, benefits, challenges, and use. Additional information was gathered using empirical methods of conversations with some respondents to provide additional information.

5.1 A snapshot of educational technologies in a Fijian classroom

The intention of the research that was conducted was to establish the status of educational technologies in some schools around Fiji. The Fijian classroom is still in the early stage of modern evolution of teaching practices. While majority of schools in Fiji are still contented with the cheap method of chalk and blackboards for delivery, many Fijian educational leaders are looking forward to modernizing their schools with educational technologies to initiate e-Learning in their school.

The survey data was reviewed with respect to the four areas involved in the adoption and use of educational technology in Fijian classroom—implementation, benefits, challenges, and uses.

5.2 Summary of research and discussion

5.2.1 Implementation

Question	Response
Does the school provide educational technologies to the teaching staff and/or students?	All respondents indicated that educational technologies were being provided to the staff. While most schools used technology as a teaching tool in a few classrooms and special rooms, one school affirmed that their entire teaching pedagogy was now based on using educational technology to deliver lessons to student in classrooms.
Please answer this question. Which of the following technologies is available in your institution?	The distribution of hardware resources available in the sample is given. 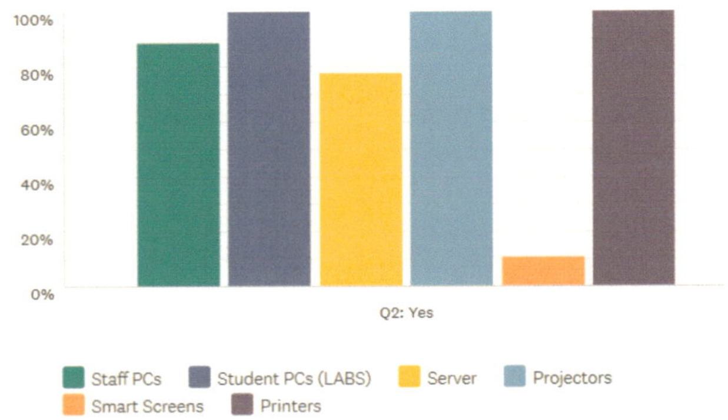

Fijian administrators have realized that educational technologies are a step forward in the Fijian classroom and are taking steps to slowly introduce educational technologies to support e-Learning in segments. They recognize the relevance of educational technologies to the Fijian classroom and are making efforts to step forward into the future.

5.2.2 Benefits

If you have answered Yes to Question 4, state the benefits from implementing such a system.	The following benefits were summarized from the responses gathered: 1. The use of technologies allows for a child-centered approach making it easier to deliver lessons to students. 2. It provides an additional learning support for students. 3. This modern, different approach from the normal board and chalk boosts interest from children allowing the teacher to capture a lot more of the content while teaching since lectures are accompanied with demonstrations easily which makes teaching effective. 4. Interesting method of learning for students since it is learner friendly. 5. With multimedia, other related information to lesson can be shown to students. 6. The use of a different teaching method available on education websites and video sharing sites when shown to the class increased the student's awareness on other learning tools available through the Internet and reinforces learning outcomes far more efficiently. 7. The students' interest continues to build up as they learn. 8. Students get engaged for longer time.

	9. Very efficient way of delivering content to the twenty-first-century students as they are known as digital natives. They are more interactive with educational technology for educational purposes. 10. Promotes discovery learning whereby students can also do presentations using technology. 11. Copies of class resources will remain conveniently organized and available with teachers and students for longer periods. This promotes flexibility since content can be edited and shared as seen fit. Makes teaching efficient. This is termed as working smart. 12. Images used and projected on the screen are clear and colorful. This increases the level of understanding on the subject matter. 13. Teachers and students remain abreast with data as technology usage increases research and analytical skills. 14. Raises the professional standard of teachers to another level.

The benefits listed are summarized from the responses received from Fijian educators. They are vividly clear on the benefits attained from the uses of such technologies in the least, delivery of lessons. The future solutions provided by the use of educational technologies remain a common identifier.

5.2.3 Challenges

If you have answered Yes to Question 4, state the challenges faced during the implementation of such a system.	1. The cost of acquiring educational technology is steep. Single implementation is not feasible in an academic year due to lack of single grant approval. Implementation over time could see technology become obsolete before completion. 2. Regular electricity outages in some areas remain a challenge. 3. Availability of experts to plan and improve architecture, purchase, implement, and train users. Maintenance and network management is quite difficult as well. 4. Preparation of electronic lessons is quite time-consuming and affects current teacher planning. Content development is also difficult for teachers as some may not be well versed with computer technology. 5. Some teachers find using educational technology quite challenging since they themselves are not trained in the use of these technologies. 6. Some students may resist the use of technology for teaching. Technology acceptance is not absolute. 7. Reluctance of management to provide technological resource since these investments requires thousands of dollars of investment. Educational technology investment across an entire school may require tens of thousands of dollars of investment. 8. If implementation depends on Internet connectivity, then downtimes and throttled speeds remain an issue. 9. Mishandling of equipment. Some students take technology for its entertainment value only. Some teachers may not be secure with their credentials allowing students to engage in abuse of resources.

Given the challenges outlined, the relevance of educational technologies to the Fijian classroom faces its key argument. These outline the present state of educational technology implementation in Fiji. These issues are real and diminish the swiftness with which educational technologies need to be introduced. These are the key areas that need to be addressed thoroughly before the relevance of educational technologies in a Fijian classroom I recognized.

5.2.4 Use

If you have answered Yes to Question 3, which educational technology for e-Learning approach has been implemented in your school?	• The most common implementation of educational technology is for technology as a teaching tool (78%) for teacher delivery of lessons using a PC, projector, and whiteboard/screen. • The very large school has either piloted (22%) the use of PC, projector, and interactive boards or implemented it completely in their schools. • The use of smart board remains non-existent in the sample. • Total e-Learning implementation with teacher delivery and student portal is not used either, and it seems that this would need more time to be accepted as it requires equal acceptance and investment from parents and other stakeholders as well.

The key fact here remains that educational technologies have limited implementation in the sample. The limit of use remains with teacher delivery only. The true potential of educational technologies in the form of e-Learning remains to be seen.

6. Findings

Based on the data analysed and literature reviewed, a comparison was made with present technologies and researched data; it is evident that some forms of educational technologies are being utilized albeit its limited use in most Fijian classrooms. It also remains non-existent in many schools but these schools were not sampled. Awareness and desire to implement and utilize educational technology remain elevated. The benefits of such implementations far outreached the challenges. The greater belief is that students will benefit vastly should teachers take care in using technology and implementing carefully despite the challenges faced.

An immense need for stakeholder participation and investment was outlined in order to realize the potential of educational technology and eventually e-Learning implementations. Fijian administrators are quite intrinsically aware of the benefits of educational technology since they have been beneficiaries of such tools in their time in universities as students and as such recognize the relevance of educational technologies in their classrooms.

Based on the feedback, educational technologies' relevance in a Fijian classroom cannot be questioned. Times have changed as students' exposure to a myriad of technology requires them to be placed in classrooms of similar nature. Students would possibly draw more benefits from such initiative since most challenges do not involve them directly instead concern other stakeholders.

7. Conclusion

To conclude, educational technologies can no longer be ignored as emerging technology since its relevance and presence to Fijian classrooms are realized by educators. Given the types of educational technologies available around the world and those that have been sourced by the sample, it is clear that most solutions are present in the Fijian classroom and have been acquired and used by some schools already. While not all technologies are readily available or affordable, research shows that Fijian classrooms are slowly investing in educational technology. While acceptance of these technologies constitute for an in-depth research, the relevance to our education system remains divided. Even though the benefits are outlined and some form of educational technology implementation has been made, the challenges faced are also real.

The cost of investment is the single primary deterrent to making educational technology and subsequently e-Learning more relevant given that not enough investment or confidence is being placed into such advancement by the key financial stakeholders. Looking at the research, the implementations are limited for each school. Only one school in the sample has a complete installation of educational technologies. The rest of the schools have very few classrooms engaged.

This simply indicates that the Fijian classroom is not quite ready to completely embrace educational technologies into their operations given the challenges are few but immense. The leap in effort from all stakeholders is imperative for educational technology and eventually e-Learning to find their relevance in a Fijian classroom.

8. Future directions

The following questions, when answered, indicate a possible future direction.

Are there any accelerated efforts for plan and preparation for educational technologies for the nation's classrooms from the key stakeholders?

Are there milestone requirements of authorities for schools to implement educational technologies in stages for the benefit of students?

Are national syllabi team planning and delivering content and tools that are compliant to educational technology standards?

Author details

Reginald Gani[1], Sharika Devi[2], Sam Goundar[3], Emmenual Reddy[3]*
and Fatemeh Saber[3]

1 The University of the South Pacific, Nausori, Fiji

2 The University of the South Pacific, Lautoka, Fiji

3 The University of the South Pacific, Suva, Fiji

*Address all correspondence to: emmenual.reddy@usp.ac.fj

References

[1] Aziz H. The 5 keys to educational technology. The Journal—Transforming Education Through Technology. 2010 Available from: https://thejournal.com/articles/2010/09/16/the-5-keys-to-educational-technology.aspx [Retrieved: 20 21, 2018]

[2] Januszewski A, Molenda M. Educational Technology: A Definition with Commentary. New York and London: Routledge; 2013

[3] Kumar S, Daniel BK, Integration of learning technologies into teaching within Fijian Polytechnic Institutions. International Journal of Educational Technology in Higher Education. 2016; **13**(1):36. Available from https://doi.org/10.1186/s41239-016-0036-8

[4] Ghavifekr S, Rosdy WA. Teaching and learning with technology: Effectiveness of ICT integration in schools. International Journal of Research in Education and Science. 2015:176

[5] Domingo MG, Gargantè AB. Exploring the use of educational technology in primary education: Teachers' perception of mobile technology learning impacts and applications' use in the classroom. Computers in Human Behavior. 2016; **56**:21-28

[6] Gosper MM, Pizzica JM, Ashford-Rowe K. Student use of technologies for learning—What has changed since 2010? In: Hegarty B, McDonald J, Loke S-K, editors. Proceedings of ASCILITE 2014. Dunedin: Australian Society for Computers in Tertiary Education; 2014. pp. 290-301

[7] Stosic L. The importance of educational technology in teaching. International Journal of Cognitive Research in Science, Engineering and Education (IJCRSEE). 2015;**3**(1)

[8] Ross SM, Morrison GR, Lowther DL. Educational technology research past and present: Balancing rigor and relevance to impact school learning. Contemporary Educational Technology. 2010;**1**(1):17-35 Available from: http://dergipark.gov.tr/cet/issue/25719/271396

[9] Selwyn N. Distrusting Educational Technology. New York: Routledge; 2013

[10] Office of Educational Technology. Reimagining the Role of Technology in Education. U.S. Department of Education; 2017

[11] Sang G, Valcke M, Van Braak J, Tondeur J. Student teachers' thinking processes and ICT integration: Predictors of prospective teaching behaviors with educational technology. Computers & Education. 2010:103-112

[12] Simon J, Garcia-Belmar A. Education and textbooks. Technology and Culture. 2018:940-950

[13] Yadav N, Gupta K, Kethrapal V. Next education: Technology transforming education. South Asian Journal of Business and Management Cases. 2018:68-77

[14] Winthrop R, Williams T, McGiveny E. Innovating to unburden teachers. In: Education and Development. Washington, USA: Brookings Institution Press; 2016

[15] Hockly N, Gavin D. Current and future digital trends in ELT. RELC Journal. 2018:164-178

[16] Tondeur J, Roblin NP, Van Braak J, Voogt J, Prestridge S, Forkosh-Baruch A, et al. Effective approaches to prepare future teachers for educational technology use. In: Proceedings of Society for Information Technology & Teacher Education International

Conference. Savannah, GA, United States: Association for the Advancement of Computing in Education (AACE); 2016. pp. 3082-3085

[17] Lund A, Furberg A, Jonas B, Engelin KL. What does professional digital competence mean in teacher education? Universitetsforlaget Nordic Journal of Digital Literacy. 2014;**9**: 281-299

[18] Blackburn G. My end is my beginning: Elearning at the crossroads. The Turkish Online Journal of Educational Technology. 2016:87-97

[19] Darling-Aduana J, Heinrich CJ. The role of teacher capacity and instructional practice in the integration of educational technology for emergent bilingual students. Computers & Education. 2018;**126**:417-432

[20] Wu SP, Corr J, Rau M. How instructors frame students' interactions with educational technologies can enhance or reduce learning with multiple representations. Computers & Education. 2019;**128**:199-213

[21] Shah T. Facilitator Technology for the Future Net. 2018. Digital Learning. Available from: http://ezproxy.usp.ac.fj/login?url=https://search-proquest-com.ezproxy.usp.ac.fj/docview/2047324208?accountid=28103 [Retrieved: August 17, 2018]

[22] Taylor II RP. The computer in school: Tutor, tool, tutee. Contemporary Issues in Technology and Teacher Education; 2013

[23] Data Projections. 4 Different Types of Educational Technology Software Available. 2018. Available from: https://www.dataprojections.com/dp-blog/4-different-types-educational-technology-software-available/ [Retrieved: October 23, 2018]

[24] Modern Consumers. Types of Classroom Technologies. 2017. Available from: http://modernconsumers.com/types-classroom-technologies/ [Retrieved: October 23, 2018]

[25] Statistics How To. Convenience Sampling (Accidental Sampling). 2015. Available from: http://www.statisticshowto.com/convenience-sampling/ [Retrieved: October 24, 2018]

Chapter 5

Learning from Online Voices: A Mixed Methods Approach to Explore Patient Online Reviews of Hospital Care in Portugal

Carla Marisa Ferreira Gomes,
Marlene Paula Castro Amorim and
Mário Jorge Ferreira Rodrigues

Abstract

Online patient reviews can offer a rich information source to users of healthcare services, as well as for hospital management and quality monitoring. Whereas in recent years the volume of online patient reviews has been consistently growing, organizations still lack standardized approaches and tools to allow for the systematic monitoring of users' online comments. Therefore, managers are lagging in the ability to make use of such data from patients' voices for improving the quality of the services provided. If organizations fail to develop the right capabilities to consider users' online reviews and feedback, they risk not only to miss important quality failure alerts, as wells as to frustrate their customers' expectations for service and attention. In this chapter, we present a qualitative analysis of patients' reviews for healthcare services in Portugal, building on a sample of data extracted from Google for the year of 2019. The chapter reports the major quality management themes addressed by hospital users in their online expressions and offers some guidelines to support a structured analysis and visualization of results from online users' word of mouth data.

Keywords: Online Ratings, electronic Word-of-Mouth, Quality management, Hospital, Quality in Health, User Generated Content

1. Introduction

The Quality of Health Services influences the dynamics of Health Institutions and determines the care provided to users, which are co-producers in a logic of interactivity. The collection of citizens' opinions about the Quality of Health Services, allows for a subjective report distinct from other quality measures (e.g., quality indicators). It is usually carried out in a structured, offline, and conventional way (e.g., satisfaction surveys) [1, 2], but globalization and the internet have brought a novel feedback source called analysis of online comments and classifications (OCC) [3, 4].

OCC are considered a form of electronic-word-of-mouth as they capture the essence of consumers' direct sharing of experiences just as the traditional word-of-

 IntechOpen

mouth. Word of mouth is one of the most pervasive forms of disseminating information about customer experiences. As a form of interpersonal influence is acknowledged as one of the more important drivers that influence customer purchase decisions, notably when the services that are being evaluated for purchase involve intangible attributes that cannot be experimented and assessed a priori. Online media has created an enormous arena of opportunity for companies and customers to share information. As such we have witnessed in recent years an explosion of voluntary testimonials from customers about their service encounters in a mode of expression that has been labeled as electronic word of mouth that are usually shared together with service ratings and classifications. Altogether electronic word of mouth and customer ratings altogether designed as OCC, offer a rich source of information for customer preservice evaluations. Moreover, they are also an enormous source of data and insights to inform the quality management function in the company. The remaining challenge is to devise structured methods to make sense of all the available – often unstructured – information. OCC are being produced at increasing rates and provide a different understanding of user satisfaction compared to traditional measures [5]. For many services, and particularly for the context of the health sector, the current debate concerns how to make this information relevant to support the decisions of users and hospital managers, notably by allowing for the development of methods of analysis that help in the identification of quality gaps [1, 2, 6, 7].

Studies have already been developed within the scope of OCC analysis in other areas of the industry (e.g., research goods, restaurants, hotels), but transposing these methodologies to health services is not straightforward. The investigations carried out in this area are international and are directed only at the user's decision and do not explore methods that assist managers in the continuous improvement of the quality of health services. Having such methods and tools available, health service managers would have a better understanding of the quality of the service provided, and thus have more informed decisions about how service quality can be improved. Moreover, its existence in the Portuguese context is unknown.

This investigative gap, added by factors such as the increase in the volume of this information and the apparent lack of structured methods or tools to systematically extract useful information, determined the development of this study and the identification of the following general objectives: (1) to explore ways to analyze and summarize large volumes of information available online generated by users in the context of health; (2) to devise ways for summarizing information and to propose methods to support the decision making of health institution's managers.

The analysis was conducted using a mixed method approach that involves bringing together qualitative and qualitative sources of data in a meaningful manner. Research about the quality of services has been characterized by the prevalence of quantitative approaches, building on data from questionnaires that aimed to capture customer perceptions about the various service attributes. OCC put on the table a mix of customer ratings and customer generated (narrative) content that were addressed in this work: ratings attributed by service consumers were mixed with qualitative analysis from the narrative content. The quantitative analysis, by itself, was also made mixing well known computer algorithms with a thorough manual content analysis. This chapter has five sections. The first section was the introduction to the topic and elaborated on the motivation and relevance of the topic; section two provides background and related work information; section three explains how the sample data was specified and collected; the fourth section presents the data extraction results and respective analysis. The chapter ends in Section 5 offering some discussion and conclusions.

2. Background and related work

In health, consumers do not have a passive role, but rather actors whose opinion about service experience can be decisive to support the performance of managers, notably for the continuous improvement of the service quality. It is important to create mechanisms that facilitate users' feedback, as it is difficult to give a voice to the user, but even more difficult is to incorporate the voice of the users in the decision-making processes of the management teams [8]. Thus, users' OCC can be a vehicle that facilitates communication between users and health institutions.

The concept of quality in health has evolved over time. Health services are considered credible or high-risk services and have characteristics and particularities that differentiate them from other services or products. They are considered heterogeneous, intangible and require a large citizen participation [5, 9–11]. It is not generally possible to assess its quality before experiencing it, fand the possibility of returns is also limited. It is only possible to correct surgical complications or reduce consequences that can be harmful for the user himself (e.g., incapacity at work), for the family and for society itself (e.g., increased health spending). The same does not happen with research goods or products (e.g., purchase of a smartphone) or in experience services (e.g., restaurant). In health services there is an information asymmetry between the user and the superior care provider compared to other services [11, 12].

Donabedian [13] proposed a model in which quality in health depends on interventions aimed at the three pillars: Structures of care provision, that is, attributes of the environment such as material and human resources, facilities and organization; Processes arising from the provision of care such as technical and interpersonal skills of health professionals; Results, understood as the reflection of health care in the user, which include rehabilitation and recovery of users, control of chronic illness, change in behaviors and lifestyles, and satisfaction with the care provided [14–16]. The Donabedian model is analogous to the division suggested by Rothenfluh and Schulz [16], which categorizes health care into characteristics of a research, experience, and credibility nature. For WHO, OECD & World Bank [17] Quality in Health is a continuous and dynamic process where care is sought at the right time, in a coordinated manner, responding to the needs and preferences of the citizen, minimizing damage, and wasting resources, seeking to increase the probability desired health outcomes. It is a complex process [6] and its measurement is a great challenge [16].

There are several models proposed for the evaluation of the quality of health services, namely the theoretical model of Donabedian [13], the model of Grönroos [18] and the model of Parasuraman [19–22]. The model by Grönroos and Parasuraman et al. they were initially developed for most services, and the Grönroos model is not as perfected as that of Parasuraman, whereas the Donabedian model was developed specifically for health services.

The abandonment of the paternalistic and hierarchical health model and the adoption of the model based on shared decision, fostered the centrality of the citizen. The growing awareness of the citizen forced organizations to take an interest and to promote Quality and continuous improvement policies in the pursuit of organizational excellence [23]. Thus, users' perceptions of service quality have become a critical component for measuring the quality of health care and services [24]. The authors describe two ways of assessing Quality in the health Services: health outcomes or Quality indicators (Order No. 5739/2015, May 29) [25]; and user experience assessment instruments [26], which subjectively reflect the quality of health care, in a unique and individual perspective, containing valuable information on distinct aspects of services.

There are several ways to listen to the user, namely classic or conventional methods and more recently, OCC have emerged. Classical methods are usually constituted by a standardized set of questions about the care received from a specific provider (health unit or professional), such as face-to-face, digitized or telephone satisfaction surveys [27]. This form of feedback implies the non-voluntary participation of the user, whose collaboration initiative comes from the health institution, which also has the responsibility of selecting the questions, mostly consisting of closed questions, and translating into a numerical result, as well as it is also the institution that determines the frequency with which it is applied [28].

OCC can represent a new way of assessing Quality in Health, understood as positive, negative, or neutral online publications and reviews, carried out by real or potential users, about a service or product [6]. They are an unsolicited way, therefore, voluntary for users to write opinions or perceptions about aspects related to health services online. Usually, this information is not standardized [26] and can only be translated into a punctuation (e.g., number of stars) and/or a free text, the wording of which does not obey any prerequisite in terms of structuring [11, 27, 29].

The use of OCC has advantages compared to traditional measures, particularly: websites are easy to access and use [1, 30] they are a form of information that promotes transparency, since data are made available online and are accessible to most people [28, 31]; they are a real-time barometer of public opinion, in a context of rapid and constant change, and allow the identification of prominent issues [11, 32]. They can be a substitute, instantaneous or almost in real time, for the analysis of the users' experience, with the possibility of using automated methodologies, which facilitate their analysis [31, 33, 34] and it is in this sense that the present study will seek to make a major contribution; they offer a convenient, safe, low-cost mechanism for organizations to hear users' voices [31, 34], being an important means of alarm to signal the deterioration in the quality of care [16] or, on the other hand, to identify successful practices [31].

On the other hand, there are several limitations that are mentioned in the literature to the use of OCC, which include: influence of other factors, in addition to the quality of care; anonymity and vulnerability to fraud [7]; the risk of not being representative of the general population [30, 31], noting that the literature refers that they are the youngest female users, with a higher degree of academic training, living in metropolitan areas [3, 35] and those who use health care (e.g. chronically ill) most frequently write this information.

Although consumer opinion websites have been in existence for more than two decades, the first study on health OCC was published in 2009 [36], and most studies were published after 2010 [33]. Studies published before 2010 used content analysis with small samples, whereas more recent studies recovered and studied a larger data set using automated technologies [33].

Most of the studies turn out to be descriptive, where the numerical classifications are analyzed with the determination of their frequency. There are other descriptive studies that analyze the narrative comments of OCC, understood as content analysis. If at an early stage, they used traditional qualitative methods to find the main categories of these comments [33], more recently used advanced techniques, such as Natural Language Processing (NLP) ("natural language processing"). They are, therefore, advanced analytical methods that allow the content analysis of thousands of narrative comments.

Research authors use diversified indicators to collect their data, such as the number of stars [7], the number of words [37], gender [38] or the total score. The

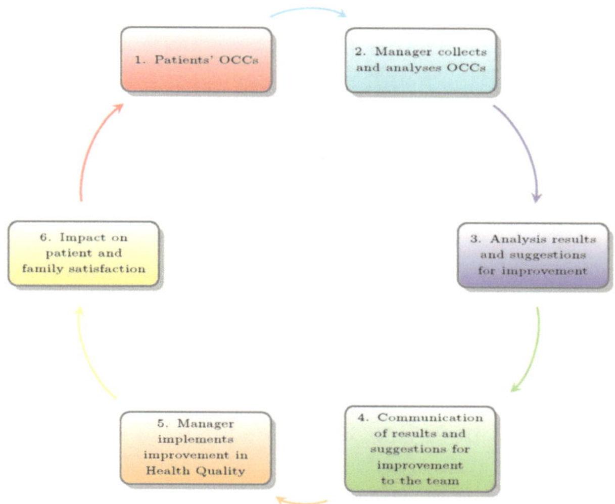

Figure 1.
Collection cycle and analysis of OCCs and their importance in the Manager's intervention.

studies that carry out content analysis may suggest a great diversity of categories within the scope of Quality in Health to group the collected data. It is also common for authors to perform sentiment analysis [39] and to classify textual data according to positive, negative, neutral and/or mixed perspectives.

OCC are increasingly important sources of information for making diverse types of decisions [40]. Understanding and systematizing them is urgent and essential for the citizen and the manager [2]. The following model (**Figure 1**) shows that a systematic collection and analysis of this information by the Manager can feed and have a direct impact on the continuous improvement of the Quality of Health Services.

3. Sample definition and data collection

The main purpose of this work is to develop and discuss methods to process the diverse and rich information that is available in the online platforms, and that is offered predominately in an unstructured format. Despite the richness, diversity and volume of this information, managers still lack structured approaches to deal with such large volumes of data, and to make sense of it for supporting decisions and initiatives for quality management. In this work we offer a contribution for the advancement in the knowledge about how to extract and categorize existing online customer generated content offering a structured approach to make sense of the data. Specifically, the research work builds on prevalent service quality models, that offer an established multidimensional approach that has validated over the year a range of distinct service quality constructs (e.g. reliability, empathy, etc.) that frame the domains of assessment that customers consider when evaluation service experiences. The study builds on such service quality conceptualization to classify the customer reviews content and develop summary metrics to inform quality management though in the context of health services. The study offers a stepwise view of the content analysis deriving highlights about critical aspects in the extraction of data, as well as in the process of cleaning the extracted data for the purpose of delimiting a meaningful sample of customer comments. The advancements in such data extraction and analysis are of critical importance, particularly for medium sized organizations, who might lack the resources to devote specific capacity to make use of the available information. Moreover, as the pace of expansion of the available customer content continues to accelerate, the need to deploy methods to analyze content that are aligned with the prevalent quality models, that are embedded in current managerial knowledge is an urgent matter.

An exploratory-descriptive, cross-sectional, and qualitative study was developed using content analysis after selecting a rational and criterial non-probabilistic sample. It seeks to respond to the following specific objectives: to identify relevant variables for the segmentation of content generated by users; identify relevant indicators that allow monitoring the behavior of this information and signal the moment when the manager must extract and analyze the content generated by the users; and to identify categories that classify the content contained in the online comments generated by citizens in the context of health, in line with the quality models in health services.

The data was extracted from Google's online platform and is freely accessible to anyone. The choice of this platform is due to accessibility, reach and the fact that it covers all services and national territory. Ethical considerations were safeguarded since the data that could identify the population or sample studied were hidden and submitted to codification, preserving anonymity and confidentiality.

The determination of the sample that was intended to be representative of the comments, was a complex and time-consuming process associated with the dispersion and volume of information, as well as the diversity and form of organization of the Portuguese Health Institutions, as shown in **Table 1**, implying continuous methodological adjustments. It included two stages, namely the delimitation of the Institutions to be included in the sampling process and the second stage, which focused on extracting comments and selecting valid comments, that is, on determining the units of analysis.

Management model	Hospitals (Hs)	Typology (portaria n° 82/ 2014)						N° of Hs Total	N° of OCC
		I	II	III	IV-a	IV - b	IV - c		
EPE	ULS	8	—	—	—	—	—	8	7735
	CH	11	5	5	—	—	—	21	
	H	5	3	—	—	—	1	9	
	I	—	—	—	3	—	—	3	
SPA	ULS	—	—	—	—	—	—	—	243
	CH	—	—	—	—	1	1	2	
	H	—	—	—	—	—	—	2*	
	I	—	—	—	—	—	—	1*	
PPP	ULS	—	—	—	—	—	—	—	888
	CH	—	—	—	—	—	—	—	
	H	3	—	—	—	—	—	3	
	I	—	—	—	—	—	—	—	
Total		27	8	5	3	1	2	49	8866

EPE: Entidades Públicas Empresariais (public business entity); SPA: Sociedade Pública Administrativa (administrative public society); PPP: Parceria Público-Privada (public-private partnership).
**Entities not discriminated by portaria n° 82/2014 (10 abril).*

Table 1.
Distribution of the number of OCC in the different hospital institutions.

In the first stage, we applied preliminary delimitation criteria (typology of Hospital Institutions (IH's), total number of OCC and IH management models) and all SPEs except Braga Hospital and group IV were selected. Of the initial 8866 OCC, 7188 OCC remained, so it was necessary to carry out an additional delimitation, using the formation of clusters according to additional and previously defined criteria (Type of HIs; Population covered (year 2018); No. of consultations (year 2018); number of urgencies (year 2018); number of surgeries (year 2018); total number of OCC), to form homogeneous groups. In this sub step, 4 Type IHs (ULS in the Northeast, the West CH, the Baixo Vouga, the ULS of the Alentejo Coast) were selected; 1 Type IHs (CHU of the Algarve); 1 Typology III HIs (CHU S. João) revealing a set of HIs that are representative of the country's diversity and accounted for a total of 1088 OCC.

In the second stage, the extraction of the OCC was done automatically, using the research team's software, to a database elaborated using Microsoft® Excel® Office 365MSO, which is organized based on previously defined indicators. The definition of the indicators to be included in the study were based on two sources: indicators used in previous investigations (narrative commentary, number of stars, gender, number of words) [7, 38] and indicators defined in the scope of this study (date, location, language, total score of the institution and number of Likes).

After a posterior extraction of 1179 OCC, which included the previous 1088 and increased due to the dynamic character of these data and platforms, a preliminary selection of comments was made eliminating obsolete data, specifically the 437 OCC that did not present characters and the 150 irrelevant OCC (with meaningless criticism or with a meaningless argument) (**Table 2**). Subsequently, for the 592 OCC included in the study, the units of analysis were defined, fragmenting the comments into different thematic categories. The thematic categories resulted from the joining of the perspective of two authors: the model of Parasuraman et al. [20] and Gillespie and Reader [41]. This combination made it possible to define four categories to be used in the present study: "Tangibility", "Response Capacity", "Empathy" and "Reliability/Guarantee". The models selected for the definition of the categories to be used in the present study are Quality models, which, although one of them is old, still presents current dimensions for the existing services, namely in the scope of health services. Thus, in the 592 OCC selected, 898 sub-comments were identified, which constituted the units of analysis of the present study.

Total of extracted OCCs: 1179					
		Excluded		Included	
Typology/Cluster	Hs	No text	Irrelevant	Relevant	After Categorization
I / 1	ULSN	21	4	15	24
I / 3	CHBV	73	20	99	111
I / 3	CHO	76	25	74	140
I / 4	ULSLA	23	9	19	36
II	CHUA	104	35	219	339
III	CHUSJ	140	57	166	248
	Total	**437**	**150**	**592**	**898**

Table 2.
Distribuition of OCC.

4. Results and analysis

The present study combined computerized automatic analysis with qualitative techniques for the analysis of feelings and the categorization of comments in the domains of Quality in Health. The relevant comments considered in the analysis were categorized as shown in **Table 2**.

The comments were mostly written in Portuguese, but 15 other languages were also found (e.g., Korean, Finnish, Slovak, Turkish, Indonesian, etc.). The translation was automatic, using Google translator, a methodological option justified by the need to facilitate the replicability of the present study. The HIs with the highest percentage (37%) of foreign comments was type II, possibly associated with the tourist flow.

The OCC were performed mainly by men, except for the I/1 ULSN typology and III CHSJ typology. The genre was obtained from the nickname associated with the comment, but some publications were made by foreign citizens with characters related to their language (e.g., Korean), which makes it impossible to automatically classify them. The remaining results will be presented according to each of the three specific objectives outlined previously in point 2.

The information contained in data was quite disorganized and unstructured, which made it difficult or impossible to analyze. It was noticed that, similarly to what happens in the platforms of other industries (e.g., TripAdvisor), its pre-segmentation would facilitate the extraction and analysis of data, the specification of which could be previously defined in pre-filled fields outlined in the online writing platforms. Available to users using health services. The sentiment analysis (negative, neutral, and positive) of the comment was obtained through a state-of-the-art platform for content analysis in Portuguese called Linguakit [42]. A prior automatic classification was carried out for each of the comments. However, the automatic analysis of the feeling of the text, in the negative, neutral, and positive categories, has not proved to be very robust because it often has wrong syntactic constructions, resulting from the context of free and informal writing in which they originate. This forced this automatic classification to be compared with a second manual classification, to guarantee the reliability and validity of the results. In this manual classification of feeling, a new dimension was introduced, called "mixed," when the comment refers to various aspects of Quality of Service or to several episodes experienced in health care, where the positive and negative feeling coexist simultaneously.

A comment was considered positive (when the editor mentions aspects of health quality that made him happy or pleased with health care), negative (when the editor mentions unpleasant aspects related to health services), neutral (when the comment has neither a positive nor a negative feeling) and a mixed feeling (when in the same comment the writer simultaneously mentions positive aspects, but also negative experiences related to health care). Subsequently, the feeling and the number of stars were related, where a negative feeling in the user's textual account, would demonstrate from the outset that he would be dissatisfied with his experience in health care and would score it with 1 or 2 stars, if the comment were neutral, would give 3 stars and if it were positive, it would give the value of 4 or 5 stars.

In summary, from the results presented, it is concluded that, it is extremely important that online publishing platforms find mechanisms that predispose the user to explain the feeling affection to the report about their experience in health care, since, this care prior facilitates the visualization, as well as the collection and treatment of the data made available about the user's opinion and therefore facilitates the management of information and the intervention of the manager, allowing him to provide himself with easily accessible information that can be useful for the improving the performance of the health organization. Since, and as our results demonstrate, treating and classifying *a posteriori* the feeling of a large volume of data published by users about health services can be a challenging task, where there is a considerable risk of bias in the data, in addition to predisposing to the use of complex methods, supplementary surveillance, as well as spending unnecessary resources and time. Therefore, the need to requalify the comments considered neutral, reinforces that, in fact, it is necessary that the platforms specify fields that fragment the information from the start so that it can be easily managed (**Table 3**). These results add to the body of knowledge in service quality as well as to the domain of information science, in particular for studies addressing electronic word of mouth, and they bring forward specific challenges in the analysis of content and advance highlights on how to design adequate platforms for the collection for customer inputs in a manner that meets the requirements of subsequent phases, – i.e. content analysis.

From the previous results, one of the difficulties that the manager may face when dealing with this data is related to the question of the reliability of the automatic analysis of the content. On the other hand, greater reliability will require manual analysis, and this involves spending additional resources (e.g., time, human resources) for management. The results of the present study suggest that instead of an automatic classification followed by a manual classification, a viable alternative would be to do an automatic analysis, to check the consistency between the number of stars and the sentiment and finally, manually analyze those comments where the sentiment is the inverse of the score. In other words, the alternative is to carry out automatic classification first, and then, only in case of incongruities, a subsequent manual classification.

4.1 Categories for the classification of content generated online by users, in line with the principles of quality in health services

One of the purposes of the present investigation was to identify categories to classify the online content generated by users. This was carried out in two ways: automatic and manual analysis of feelings as developed in point I. and the categorization of comments in the domains related to Quality in Health, previously defined:

Classification	Feeling	1 star	2 stars	3 stars	4 stars	5 stars
Automatic	Negative	49	3	0	3	3
	Neutral	13	2	5	2	2
	Positive	12	1	3	4	17
Manual	Negative	68	6	2	1	1
	Neutral	2	0	3	3	1
	Mixed	2	0	3	0	1
	Positive	2	0	0	5	19
Manual – Automatic	Negative	19	3	2	−2	-2
	Neutral	−11	−2	−2	1	−1
	Mixed	2	0	3	0	1
	Positive	−10	−1	−3	1	2

Table 3.
Process used to determine the differential between automatic classification and manual classification for type 1, cluster 3 CHBV.

"Tangibility", "Response Capacity", "Empathy" and "Reliability" /Guarantee". For this last categorization, an automatic classification was attempted, which proved to be inefficient, due to the characteristics of the text, since it presented some peculiarities, namely sarcasm, irony, spelling errors, acronyms, abbreviations, and stories. Therefore, the author classified manually in the four categories and one of the elements of the research team made an independent review, and in the case of disagreement, the categorization was discussed and resolved between the two elements.

4.1.1 Categorization of sentiment analysis

The results show that in relation to the sentiment, in the total of the 6 IHs it was found that 65% of the comments were negative (383/592), 1% neutral (6/592), 8% mixed (48/592) and 26% positive (155/592). In literature was found that there is no consensus, and the present investigation is in line with the results found in the study by Emmert et al. [39].

Emmert et al. [38] refer that 80% of all comments (average length of 45.3 words $\pm$42.8) were classified as positive, 4% as neutral and 16% as negative. And that longer narrative comments were more likely to be negative, while shorter comments were more likely to be positive. When an association was made between the sentiment and the size of the OCC (measured by automatically counting the number of words in each comment) it was found that the average dimension (in number of words) of the positive comments (43.98 words) differs from the negative ones (61.69 words) for the analyzed sample (t test, Sig. 0.000). Thus, we verified through the t test that the negative comments were longer than the positive ones, in accordance with the literature and similarly to the results obtained in the studies by Rastegar-Mojarad, et al. [37] and Emmert et al. [39]. Information-rich analyzes tend to be longer, with a consequent increase in utility for readers [43] and even though longer narratives were more likely to be negative [28]. In this way, it is understood that the longer comments say more and tend to be mostly negative. Thus, the results of the present investigation show that the manager must be aware of the behavior of this indicator when analyzing online data.

It was also found that users put more likes in negative comments. The results show that there are differences between the number of likes of positive comments (average 0.95) and negative ones (average 1.41) (t test, Sig. 0.000). When the association between sentiment and the average score in stars was made, it is not possible to infer that the number of stars in the positive comments differed from the negative ones (t-test, Sig. 0.144). In particular, the average number of stars for positive comments was 4.60, for the sample analyzed, and the number of negative comments was 1.31.

4.1.2 Rating of comments regarding the quality of health services

Understanding what topics are most often spoken by the user's "voice" can help caregivers, managers, and administrators of health institutions to improve the user-centered health system. This type of content analysis provides the manager with richer information that can support the development of improvement actions.

All sub comments were liable to be classified under an attribute of Quality in Health, as can be seen in **Table 4**, with none remaining unclassified. But there is a wide diversity of results, an irregular pattern that made it difficult to appreciate and, therefore, the establishment of associations.

The most debated topics in users' online newsrooms are about the "Response Capacity" dimensions (for 3 of the HIs - I / 3 CHBV typology, I /3 CHO typology

Typology/Cluster	IH	Tangibility			Responsiveness			Empathy			Reliability/Warranty		
		n	%	Average PE	n	%	Average PE	n	%	Average PE	n	%	Average PE
I / 1	ULSN	4	17%	3,0	6	25%	1,8	6	25%	2,3	8	33%	3,0
I / 3	CHBV	11	8%	1,8	60	43%	1,4	32	23%	2,4	37	26%	2,1
I / 3	CHO	20	18%	2,3	43	39%	1,9	26	23%	2,5	22	20%	2,8
I / 4	ULSLA	4	11%	1,8	8	22%	1,5	8	22%	2,6	16	44%	2,8
II	CHUA	37	11%	2,2	112	33%	2,0	70	21%	3,0	120	35%	2,6
III	CHUSJ	31	13%	3,2	102	41%	1,7	41	17%	2,7	74	30%	2,9

Subtitles: IH = Hospital; PE = Star Rating; n = total ORCs.

Table 4.
Distribution of comments by average star ratings.

and III CHUSJ typology) and "Reliability /guarantee" in the remaining 3 IHs (typology I /1 ULSN, typology I/4 ULSLA and typology II CHUA). If, on the one hand, the literature stresses that, the user may not have the competence or knowledge capable of evaluating the technical or clinical aspects of care, that is, aspects related to the "Reliability / Guarantee" dimension and that its evaluation may be inaccurate [16, 21], on the other hand, in this study it appears to be an aspect in which users focus their attention, as can be seen for the values related to the dimension "Reliability/Guarantee".

"Tangibility" was the least mentioned dimension for all the HIs in the studied sample, being, therefore, the subject least mentioned by users (**Table 4**). It was observed that the aspects of "Tangibility" are the least highlighted in the comments available to all IHs by users, compared to the other dimensions of QeS, so that a change in this proportion, that is, the increase in the percentage of tangible aspects in view of the other domains, it could be an alarm signal that is the target of investigation and intervention by the manager of the health organization. Thus, and given the national context where most public health institutions were built several years ago, resulting from the triggering of the NHS, and that the Portuguese historical and economic context, as well as the financial crisis that hit Europe and that had repercussions in the divestment in the facilities and material resources of public services, may be an explanatory factor for users to have tolerance regarding the tangible aspects existing in the services of the HIs. Not least because, much of the investment made in the health sector was directed to flagrant problems such as the increase in the prevalence of chronic diseases resulting from the increase in average life expectancy [44]. It is a fact that, it seems natural that users have elevated expectations about the functioning of public health services, as these are financed by their taxes, but their attention is focused on aspects related to "Capacity response" and "Reliability/Guarantee" and not so much for the characteristics related to the appearance of the facilities, professionals, and equipment. Health services are a type of service that is not visible to everyone because they are characterized by an asymmetry of information between the user and health professionals [11]. In addition, these services are produced while they are consumed, and therefore have an intangible and heterogeneous character [5, 10, 11].

The "Response Capacity" seems to be the dimension to which users are less satisfied in this investigation since there is a lower average score value in stars for all the target HIs studied. And there was no average score in the number of stars above 3.2, it appears that the degree of user satisfaction with the selected HIs will not be high.

5. Conclusions and future work

The volume and diversity of information available online about health services implies challenges in the analysis, scarce methods or tools that facilitate its use by the manager in the continuous improvement of the Quality of Health Services. Health systems must, increasingly, be attentive to the opinions of the users who are the ones who experience the care, arranging economic tools that allow to listen to their perception and to effectively profit from the preciousness of these data. Since the standard tools for extracting and analyzing this information are not adapted to the syntactic characteristics of users' newsrooms.

The present study intended, on the one hand, to contribute to fill the investigative gap present in the national community, as well as to find effective and facilitating ways of making the most of these data, in a health system marked by the scarcity of resources. This scarcity of resources ended up being exponentiated by the current context of the pandemic, and the real impact of the pandemic on the health system, on public administration, on society in general and on each one of us has yet to be investigated.

The study allowed the identification of relevant variables for segmenting the content generated by users: hospital episode, editor, service, professional and the feeling of the comment (e.g., positive, negative), as the results showed that the automatic classification (positive/negative) differs significantly manual classification, demonstrating the complexity of its *a posteriori* classification implying that it is important not to compromise the voluntary character and the trust of the user. And finally, it allowed to classify the content according to attributes of Quality in Health: the comments were mostly negative (65%) and the average dimension of the positive ones (43.98) differs from the negative ones (61.69), resembling the results of previous studies [38, 39]. There are significant differences between the number of likes of positive (0.95) and negative (1.41) comments and the dimensions with the most comments were "Responsiveness" and "Reliability/Guarantee;" the least commented dimension was "Tangibility" for all HIs. Overall, one of the key contributions of the study is to advance in the classification of available customer generated content in a manner that is aligned with prevalent service quality models, i.e. establishing a correspondence between customer reviews and prevalent service quality attributes. Such approach offers manager an expedite way to make sense of the volumes of voluntary customer generated content, building on existing approaches to display service quality data. A ability to develop capabilities to make sense of the existing customer narratives is very important in a context where the volume of available context grows every day, and the risks of obsolescence of such knowledge are real. By using the service quality models as a steppingstone, this study proposes an approach to foster the development of rapid content analysis routines in service contexts. The methodology adopted in this study is replicable and able to be extended to other domains of service management where the volume of customer generated content is also gaining space, including hospitality services, education, public services to name a few. By refining the classification grid with the prevalent service quality attributes for each of these sectors, that are also already explored in the literature, the process of data extraction and classification is of straightforward application.

It is possible to identify some limitations in the present study, some of which have already been described by other researchers, namely, the fact that it contemplates data from only one platform (Google) [43, 45], with the risk of non-representativeness of the general population. Moreover, the study is focused on the exploration of customer content for the specific domain of health services, therefore leaving out particularities of other service sectors (e.g. hospitality, education, etc.). Nevertheless, and as explained in the results section, the method applied can be replicated in such sectors given that there is a necessary and preliminary revision of the service attributes used for the classification of customer reviews and content to match those that in the literature, have been identified as pertinent for each case.

In terms of future research, the following fields of intervention are suggested: development of software capable of processing and classifying data produced by users, adapted to their style of writing; explore, develop and adapt methodologies to other health institutions (e.g. health centers, long-term care units and private health units); and elaborate a project applicable to hospital units, allowing the results of the study to be transferred from the paper to measures and proposals for effective and practical improvement in health institutions. As mentioned by Hong et al. [33] the results of the investigations should go beyond simple descriptive analysis and theory-based hypothesis testing to provide more clinical and political implications.

Acknowledgements

This work was financially supported by the research unit on Governance, Competitiveness and Public Policy (UIDB/04058/2020) + (UIDP/04058/2020), funded by national funds through FCT - Fundação para a Ciência e a Tecnologia.

Author details

Carla Marisa Ferreira Gomes[1], Marlene Paula Castro Amorim[2]* and Mário Jorge Ferreira Rodrigues[3]

1 Rehabilitation Nurse, CHBV, Aveiro, Portugal

2 GOVCOPP and DEGEIT University of Aveiro, Aveiro, Portugal

3 IEETA and ESTGA University of Aveiro, Aveiro, Portugal

*Address all correspondence to: mamorim@ua.pt

References

[1] Huete-Alcocer N. A Literature Review of Word of Mouth and Electronic Word of Mouth: Implications for Consumer Behavior. Frontiers in Psychology. 2017;8(1256):1-4.

[2] Amorim M, Rodrigues M. Building on eWOM to Understand Service Quality in Hotel Services. Open access peer-reviewed chapter: Book Citation Index. 2017.

[3] Boylan AM, Williams V, Powell J. Online patient feedback: a scoping review and stakeholder consultation to guide health policy. Journal of Health Services Research and Policy. 2019;25 (2):1-10.

[4] Greaves F, Millett C, Nuki P. England's Experience Incorporating "anecdotal" reports from consumers into their National Reporting System: Lessons for the United States of what to do or not to do? Medical Care Research and Review Supplement. 2014;71(5): 65S–80S.

[5] Gu D, Yang X, Li X, Jain H, Liang C. Understanding the role of Mobile internet-based health services on patient satisfaction and word-of-mouth. International Journal of Environmental Research and Public Health. 2018;15: 1-23

[6] Filieri R. What makes online reviews helpful? A diagnosticity-adoption framework to explain informational and normative influences in e-WOM. Journal of Business Research. 2015;68 (6):1261–1270.

[7] Kilaru A, Meisel Z, Paciotti B, Ha Y, Smith R, Ranard B & Merchant R. What do patients say about emergency departments in online reviews? A qualitative study. BMJ Qual Saf. 2016;25: 14–24.

[8] Donabedian A. Evaluating the Quality of Medical Care. The Milbank Quarterly. 2005:83(4);691–729. Reprinted from The Milbank Memorial Fund Quarterly (1966), 44 (3), 166–203.

[9] Argan M. Investigating word-of-mouth (WOM) factors influencing patients' physician choice and satisfaction. International Journal of Medical Research & Health Sciences. 2016;5(1):191-198.

[10] Martin S. Word-of-mouth in the health care sector: a literature analysis of the current state of research and future perspectives. Int Rev Public Nonprofit Mark. 2017;14:35–56.

[11] Shah AM, Yan X, Shah SAA, Shah SJ, Mamirkulova G. Exploring the impact of online information signals in leveraging the economic returns of physicians. Journal of Biomedical Informatics. 2019;98:103-272.

[12] Lu N, Wu H. Exploring the impact of word-of-mouth about Physicians' service quality on patient choice based on online health communities. BMC Medical Informatics and Decision Making. 2016;16(1):1–10.

[13] Donabedian A. The Quality of Care. JAMA. 1988;260(12):1743-1748. doi: 10.1001/jama.1988.03410120089033

[14] Direção Geral de Saúde (2013). Plano Nacional da Saúde 2012-2016. Maio 2013, 1-110.

[15] Donabedian A. Evaluating the Quality of Medical Care. The Milbank Quarterly. 2005:83(4);691–729. Reprinted from The Milbank Memorial Fund Quarterly (1966), 44 (3), 166–203.

[16] Rothenfluh F, Schulz PJ. Physician rating websites: What Aspects are important to identify a good doctor, and are patients capable of assessing them? A mixed-methods approach including physicians' and health care consumers'

perspectives. Journal of Medical Internet Research. 2017;19(5):1–13.

[17] World Health Organization, OCDE & World Bank (2018). Delivering quality health services: a global imperative for universal health coverage.

[18] Grönroos C. A service quality model and its marketing implications. European Journal of Marketing. 1984;18 (4):36–44.

[19] Parasuraman A, Zeithaml VA, Berry LL. A Conceptual Model of Service Quality and Its Implications for Future Research. Journal of Marketing. 1985;49 (4):41-50.

[20] Parasuraman A, Zeithaml VA, Berry LL. SERVQUAL: A Multiple-Item Scale for Measuring Consumer Perceptions of Service Quality. Journal of Retailing. 1988;64(1):12-40.

[21] Tinoco PCR, Cândido CJF, Feliciano IMPL. A importância da avaliação da qualidade percebida em radiologia e o impacto dos fatores sociodemográficos num hospital público português. Port J Public Health. 2018;1-12.

[22] Cao X, Liu Y, Zhu Z, Hu J, Chen X. Online selection of a physician by patients: Empirical study from elaboration likelihood perspective, Computers in Human Behavior. Computers in Human Behavior. 2017;37 (C):403-412.

[23] Silva SC. Avaliação da qualidade de serviço numa organização de saúde privada em Lisboa [thesis]. Instituto Universitário de Lisboa; 2014.

[24] Akdere M, Top M, Tekingündüz S. Examining patient perceptions of service quality in Turkish hospitals: The SERVPERF model. Total Quality Management and Business Excellence. 2020;31(3–4):342–352.

[25] Despacho n.º 5739/2015 Diário da República, 2.ª série — N.º 104 — 29 de maio de 2015 Retrieved from: https://dre.pt/application/conteudo/67344509

[26] Greaves F, Ramirez-Cano D, Millett C, Darzi A, Donaldson L. Use of Sentiment Analysis for Capturing Patient Experience From Free-Text Comments Posted Online. Journal of Medicine Internet Research. 2013a;15 (11):e239:1-9.

[27] James T, Calderon E, Cook D. Exploring patient perceptions of healthcare service quality through analysis of unstructured feedback. Expert Systems with Applications. 2017; 71:479-492.

[28] Boylan AM, Turk A, van Velthoven MH, et al. Online patient feedback as a measure of quality in primary care: a multimethod study using correlation and qualitative analysis. BMJ Open. 2020;10:1-10.

[29] Emmert M, Meszmer N, Sander U. Do health care providers use online patient ratings to improve the quality of care? Results from an online-based cross-sectional study. Journal of Medical Internet Research, 2016;18(9):1–14.

[30] Emmert M, Halling F, Meier F. Evaluations of Dentists on a German Physician Rating Website: An Analysis of the Ratings. J Med Internet Res. 2015; 17(1):1-10.

[31] Kowalski R. Patients' written reviews as a resource for public healthcare management in England. Procedia Computer Science. 2017;113: 545–550.

[32] Wong C, Sap M, Schwartz A, Town R, Baker T, Ungar L, Merchant R. Twitter Sentiment Predicts Affordable Care Act Marketplace Enrollment. J Med Internet Res. 2015;17(2):1-8.

[33] Hong Y, Liang C, Radcliff T, Wigfall L, Street R. What Do Patients Say About

Doctors Online? A Systematic Review of Studies on Patient Online Reviews. Journal of Medical Internet Research. 2019; 21(4):1-14.

[34] Yang H, Guo X, Wu T, Ju X. Exploring the effects of patient-generated and system-generated information on patients' online search, evaluation and decision. Electronic Commerce Research & Applications. 2015;14(3):192-203.

[35] Patel S, Cain R, Neailey K, Hooberman L. General practitioners' concerns about online patient feedback: Findings from a descriptive exploratory qualitative study in England. Journal of Medical Internet Research. 2015;17(12): 1-15.

[36] Black E, Thompson L, Saliba H, Dawson K, Black N. An analysis of healthcare providers' online Ratings. Informatics in Primary Care. 2009;17: 249–253.

[37] Rastegar-Mojarad M, Ye Z, Wall D, Murali N, Lin S. Collecting and Analyzing Patient Experiences of Health Care From Social Media. JMIR Res Protoc. 2015;4(3):1-19.

[38] Emmert M, Meier F, Heider A, Dürr C, Sander U. What do patients say about their physicians? An analysis of3000 narrative comments posted on a German physician rating website. Health Policy. 2014;118:66–73.

[39] Emmert M, Meszmer N, Schlesinge M. A cross-sectional study assessing the association between online ratings and clinical quality of care measures for US hospitals: results from an observational study. BMC Health Services Research. 2018;18(82).

[40] Grabner-Kräuter S, Waiguny M. Insights into the impact of online physician reviews on patients' decision making: Randomized experiment. Journal of Medicine Internet Research. 2015;17(4):1-16. DOI: 10.2196/jmir.3991

[41] Gillespie A, Reader TW. The healthcare complaints analysis tool: Development and reliability testing of a method for service monitoring and organisational learning. BMJ Qual Saf. 2016;25:937–946.

[42] Multilingual Open Information Extraction 17th Portuguese Conference on Artificial Intelligence, EPIA 2015, Coimbra, Portugal, September 8-11, 2015. Proceedings, 2015. Retrieved from: https://citius.usc.es/transferencia/software/linguakit

[43] Lockie M, Waiguny MKJ, Grabner-Kräuter S. How style, information depth and textual characteristics influence the usefulness of general practitioners' reviews. Australasian Marketing Journal. 2015;23(3): 168–178.

[44] Bernardino M. Gestão em Saúde – Organização Interna dos Serviços. Coimbra: Edições Almedina; 2019, 243 p.

[45] Kirkpatrick W, Abboudi J, Kim N, Medina J, Maltenfor M, Seigerman D, et al. Beredjiklian, MD. An assessment of online reviews of hand surgeons. Arch Bone Jt Surg. 2017;5(3):139-144.

Chapter 6

Development of Support System Modeled on Robot Suit HAL for Personalized Education and Learning

Keiko Tsujioka

Abstract

The purpose of this paper is to maintain quality education for each student who has her/his own differentiation and needs. To achieve it, we have developed our support system for education and learning. We have examined this support system, whether it improved or did not improve the performance of 98 students in 2015, compared with that in 2014. The results have shown to be more significant than those of the previous year. Moreover, we have observed that students' behavior toward projects have been becoming greater autonomy and positive attitude in practical class. From those results, we might be able to say that this support system works effectively in personalized education and learning (PEL) by using big data processing.

Keywords: personalized education and learning support system (PELS), feedforward control, feedback control, interactive communication, behavior patterns, big data processing

1. Introduction

In this decade, personalized education and learning has become noticeable as an educational reform by using ICT in many countries [1, 2]. Personalized education and learning are instructions designed for students, meeting their needs and matching their traits, such as their learning styles and abilities [3]. Their needs and traits are different from each other, which is so-called differentiation. It is considered that the differentiation is mainly affected by personality, cognitive style, and so on [4]. Instructors will have to teach students individually, optimizing their needs and traits [5].

There have been reported studies on personalized education and learning with computer-based e-learning system collaboratively with artificial neural network [6, 7]. Recently, however, blended learning, which incorporates both face-to-face learning and e-learning by using ICT [8–11], has begun to be reconsidered on collaborative learning from social constructivism approach [12–15].

It means, however, that instructors do not always teach each student separately but in groups at the same time. It seems hard and complicated for instructors to match with each student's trait in group. Instructors will also have to design instructions in detail so that they can optimize members in group by analyzing students' traits [16, 17].

When they analyze students' differentiation of their traits, instructors will have to prepare students' various data for analysis [18, 19]. If ICT is used in class, it would be easier for instructors to collect data of students' various kinds of data; for example, when student responds to subjects given or assigned, instructors would be able to collect students' data in real time [20, 21].

How responses have been related to subjects, for example, what students have judged and decided, or which they have chosen their responses, will be recorded and accumulated by using ICT. Instructors will be able to understand students' psychological state and predict their learning behavior from their various kinds of data, not only in class but also before and after classes. In this viewpoint, those various data, which are collected and analyzed for educational aims, will be called big data processing [22–26].

On the other hand, however, it might have been becoming more difficult for instructors because nonverbal communication must be limited when they use computers in class. It means that there might be fewer information for instructors to obtain data for analysis of learners' psychological state by observing their behavior from outside while they are using computers in class, especially, for the first year students. Along with those reasons, it seems more difficult for instructors and students to keep educational quality in collaborative learning than before without big data processing.

To solve those problems, in our research, we are proposing some methods of personalized education and learning from the systems approach in the field of educational technology [27]. Practically, we have developed our support system for personalized education and learning (PELS) [19]. This system is modeled on Robot Suit Hybrid Assistive Limb (HAL) in cybernetics theory. As an examination of this support system, we will show a practical method of education with Start-Plan-Do-Check-Act (SPDCA) cycle [28, 29].

2. Concept of system

2.1 Structural elements

In the field of educational technology, systems are combined with interdisciplinary approaches [23], especially based on social constructivism. For example, educational system, biological system, and engineering system are elements of a social system. Because they are functioning interactively with each other, their relationship must be strong.

Cybernetics claims that biology and mechanical device are mutual theories, concerning communication and control. Along with unification of both systems, it is insisted that the new combination of science and technology becomes another development. Sankai has proposed the cybernetics theory [30–33], which is a technological concept of the support system by interactive communication between machine and human or brain. This cybernetics theory is combined with cybernetics, mechatronics, and informatics (**Figure 1**, left (1)) [34]. And Sankai has developed Robot Suit HAL. From those interdisciplinary viewpoints, in the field of educational technology, there are a lot of mutual concepts of our support system. Then we have developed our support system theory, modeled on Robot Suit HAL [35].

2.2 Schematic diagram

Our concept of PELS modeled on HAL is an interactive robot suit that enables people to expand human ability. HAL has been developed for patients who have difficulties to walk, assisting their motion and instructing between the doctor

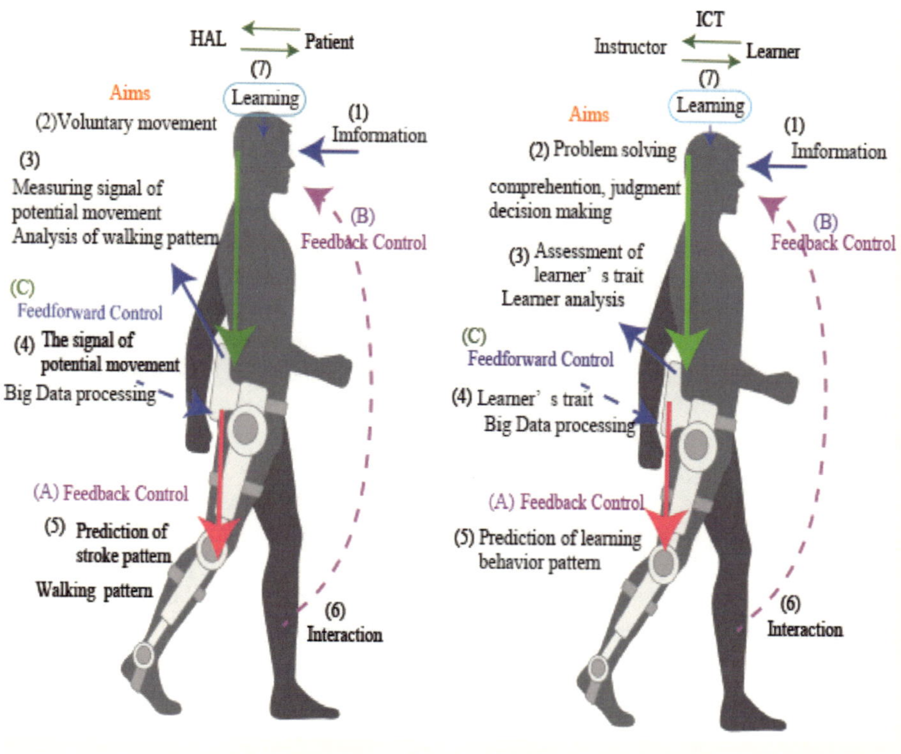

	Item	HAL⇄ Patient	Instructor(ICT)⇄Leaner
(A)	Feedback Control	HAL → Patient	Instructor(ICT) →Leaner
(B)	Feedback Control	Patient(Body)→Patient(Brain)	Learner(front lobe)→Leaner(hippocampus)
(C)	Feedforward Control	Measurement Instrument → HAL	Psychological Measurement →Instructor)
(1)	Information	Situation	Language
(2)	Aims	Voluntary Movement	Problem-solving, Judgement, Decision Making
(3)	Measurement, Analysis	Walking Pattern	Learning Behavior Pattern
(4)	Big Data Processing	Signal of Electronic Potential Motion	Learner's Trait, Characteristic
(5)	Estimate , Predict	Walking Pattern	Learning Behavior Pattern
(6)	Interaction	HAL⇄ Patient	Instructor(ICT)⇄Learner
(7)	Learning	Method of Walking, Rehabilitation	Method of Information Processing, Means and comprehension

Figure 1.
Concept of personalized education and learning support system (PELS) (revised Figure 1 in Tsujioka [36]).

and patient in rehabilitation [37–40]. HAL is composed of cybernetics voluntary control (CVC) (**Figure 1**, left (2)), cybernetics autonomous control (CAC) (**Figure 1**, left (3)), and cybernetics impedance control (CIC) (**Figure 1**, left (4)) [34] without heaviness. The wearer becomes aware of physical information by CIC of HAL. And she/he is able to obtain walking motions based on voluntary movement [41].

This system supports patients for correct walking patterns without errors (**Figure 1**, left (5)). It is accomplished by wearing HAL and comparing multiple brain areas and muscle-joint movements in real time (**Figure 1**, left (6)). Because HAL makes the brain active and the motor phenomena to reiterate correctly, its neural plasticity forward movement program is expected to improve learning effects for patients (**Figure 1**, left (7)). It is based on Sankai's hypothesis of interactive biofeedback [38, 41].

HAL consists of frames, actuators, sensors, and the main controller [42, 43], big data processing, for analysis. It is important for patients because they have different troubles with walking. Therefore, the measuring system of sensors gathers each patient's voluntary movement so that they can help patients.

HAL has two kinds of control systems, feedforward and feedback. And the main controller has been stored various data [39], for example, walking behavior and muscle movements of the unaffected leg of patients (**Figure 1**, left (C) feedforward control). Those data added signals of patients' voluntary movement from the sensing system in real time and will analyze and predict optimum walking patterns so that patients would be able to walk smoothly (**Figure 1**, left (A) feedback control). Those ideal patterns of walking will also give feedback to patients while they are getting their rehabilitation, so that it can help patients walk independently with both legs (**Figure 1**, left (B) feedback control) [34].

In the personalized education and learning support system [36] case, it has been storing various data, for example, student's traits, needs, and so on (big data processing). Instructors would be able to analyze students' data and predict students' behavior (**Figure 1**, right (C) feedforward control). And they will be able to design their instructions to meet with students' differentiation and needs, like optimized grouping members for project teams (**Figure 1**, right (A) feedback control). Through learning with those team members and matching instructions with their needs or requirements, students might also be able to know how to make collaborative relationship with each other (**Figure 1**, right (B) feedback control) [36].

2.3 Element of the system and relationship

The table in **Figure 1** [36] shows how the elements of the educational and learning support system are related to HAL. In the case of PELS system, however, language information (**Figure 1**, right (1)) should be more important for interactive communication among students in teams (**Figure 1**, right (6)). In this point, the main controller would store more various kinds of data, for example, students' performance records, differentiation, attitude, and so on, in class (**Figure 1**, right (4)).

In education, learners obtain information from instructors at first (**Figure 1**, right (1)). In this case, communication media is mainly language. After they have obtained language information about the projects, learners begin to solve problems, which are given or assigned, by judging or deciding from those information (**Figure 1**, right (2)). Instructors should predict how learners would like to make their decision (**Figure 1**, right (3)) and how they would behave (**Figure 1**, right (4)), by analyzing learners' data (**Figure 1**, right (5)), which have been collected beforehand. Because instructors need to give their feedback to students smoothly, depending on learners' traits with interactive communication, like walking with two legs normally (**Figure 1**, right (6)).

Instructors will have to decide the members of each team by some kinds of rules, so that learners will be able to communicate with each other smoothly while they are addressing problems for their projects. When learners would like to exchange their opinions or information to solve problems for given subjects, learners must be required to have some social skills (**Figure 1**, right (7)).

2.4 Hypothesis

Considering the transition of learning, it is predicted that it is more difficult for instructors to understand learners in the case of (2) or (3) than (1) in **Figure 2**. And also, in the case of learning with social media system collaboratively (4) [19],

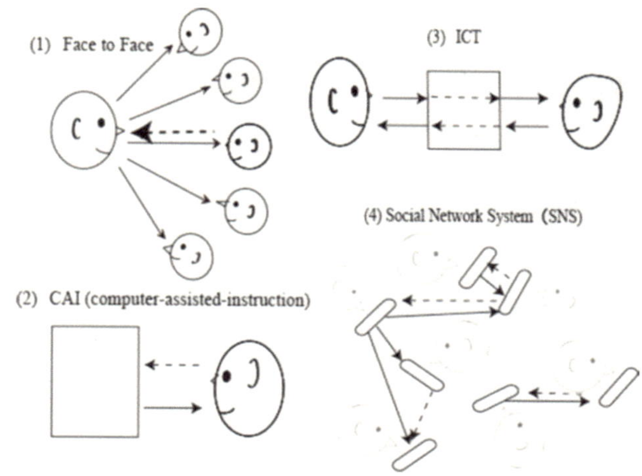

Figure 2.
Transition of education and learning configuration.

it might have been more difficult to communicate interactively, not only between instructors and learners but also among learners.

This means that it might not be easy for instructors to design instructions for learners without their learning traits which includes their learning style, personality, information processing traits, and so on (so-called big data processing), especially for new students.

The big data processing system (**Figure 3**) [19] provides learners' information beforehand with the measuring system and the assessment output system of the data analysis system so that instructors can plan their class.

Then, we have formulated hypotheses suggested by PELS model:

a. Feedforward control system (**Figure 1** (C), **Figure 4**(11)) [19] will enable instructors to predict their students' behavior (**Figure 1** (A), **Figure 4** (7)).

b. It will bring mutual understanding between students and instructors or students and students by interactive communication (**Figure 1** (6)). It might

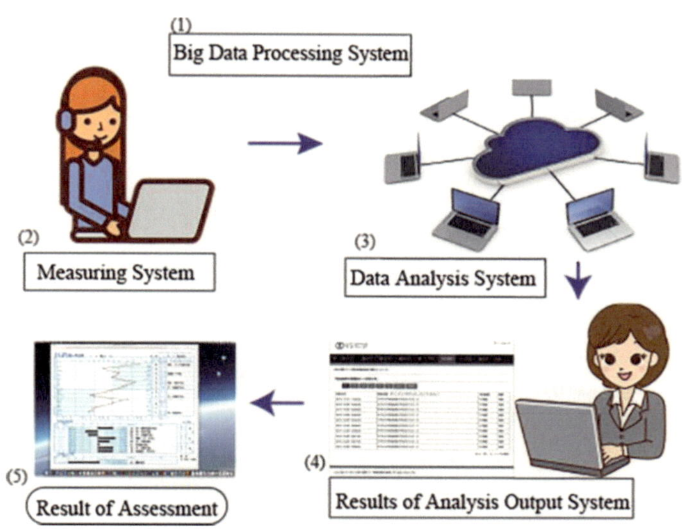

Figure 3.
Concept of big data processing [19].

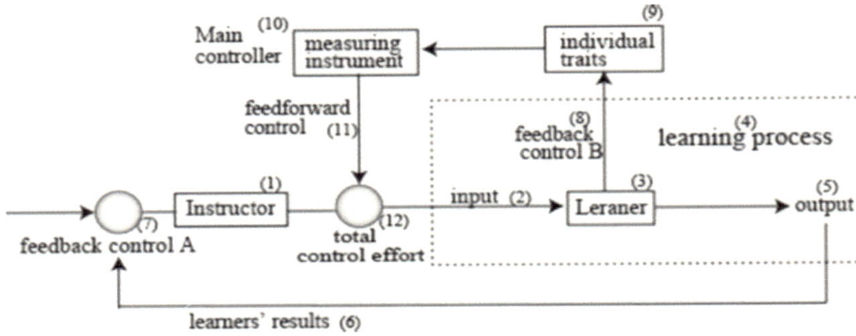

Figure 4.
Model of personalized education and learning support system [19].

help students to become independent and positive in class. Consequently, it will have effects on students' performance and attitude in class.

As those hypotheses are tested based on this PDLS model, we have designed the practical experiment along with SPCDA method [44].

3. Design

According to the theory, we have carried out a practical experiment of the PELS system to prove its effects. By using the PELS system, we will evaluate the method of education so that they might be able to improve their instruction to students, approached by SPDCA cycle (**Figure 5**, SPDCA ①-⑦) [29] based on PELS model (**Figures 1** and **2**) in order to examine the learning effect by our support system.

When we designed this plan, we also have discussed on how to examine this PELS system or whether it might work practically well, or not. At first, we had heard from a person who was participating in practical examination about what were the problems in her class, so that we can find their solution. She was a chief instructor in practical nursing class at the nursing department of the university. Her students will be eligible for the national examination after graduation to obtain qualifications of a nurse. In this reason, they are required to accumulate credits

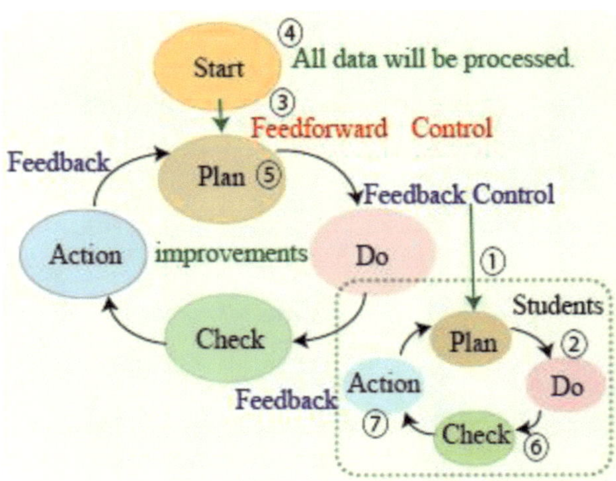

Figure 5.
The method of testing hypotheses (revised Figure 10.5 in Dannelle [44]).

toward their degrees which they will earn by standard evaluation strictly. The main problems which she has emphasized were as follows:

- A lot of students have dropped out.
- Records of students' performance have been declining.

Her aim is for students to obtain nursing concept in the first term, changing their views from general to expert. However, she thinks that it would be difficult for students to begin to be aware of the nursing expert concept. She believes that it might be related to the high ratio of leaving their job or dropping out while they are students.

Secondly, after hearing about the present situation from her, we have analyzed the problems in her class so that we could find the solution. Through finding out the reason why those problems had happened to students, we have to decide the best. Another instructor who worked at the same department said that a lot of students have been late in class and hesitated to join or start the practice for their project with team members (**Table 1**, (a) and (b) in 2014). Those phenomena might be the same problems which were mentioned by both instructors.

Finally, from those viewpoints, we became to have research questions: why students cannot study enthusiastically or what is happening to the students. Those

Points of Examination		2014	2015
(a)	Standby	a few students	most of all students
(b)	Warm up	The class has always been late about 10 minutes.	The classes has always started in time.
(c)	Attitude to the Problem Solving for Task	It was difficult to cope with a problem for practices.	After the instruction, students have began to talk with each other in order to address a problem and tackle with practice very smoothly.
(d)	Exchange of Views	There has not happened in some teams to interact with each other smoothly.	All teams took interactive communication smoothly for an exchange of their views.
(e)	Participation in Class	Some students were not able to participate in class.	Almost all students have participated in class independently.
(f)	Instructors' Reflection	No particular support was conducted. They looked for students from a subjectivity point of view.	Instructors reported that they have predicted learning behavior of students who were concerned with their attitude, referring to the consultation of psychological assessments. Instructors have also reported that they have been able to view student behavior objectively.
(g)	Grouping (each team is four members, 25 teams)	In class, the grouping of team members was organized by the order of students' ID number. Teams were not divided into groups by each instructors. In pilot experiment, the chief instructor decided to put students into groups optimizing by her experiences.	The chief instructor decided the grouping of team members, optimizing by her experiences based of assessments of psychological testing and other data. She also divided teams into five groups by each instructor's experience and the condition of an employment .

Table 1.
Comparison of attitudes toward class between 2014 and 2015.

problems should be analyzed for us to understand their present situation. We have also heard from the other instructors about the problems in 2014 (**Table 1** (c)–(e)). Bringing their opinions altogether, it is considered that their learning has been organized or formed by interactive communication between learner-learner or learner-instructor. We predicted that if their communications are exchanged smoothly with each other, those problems will be solved.

4. Solution

After we have heard from instructors about their problems in class, we have designed the practical SPDCA cycle (**Figure 2**) with PELS system based on the theory of HAL model.

1. **The first cycle** (pilot experiment)

    a. Plan

    - Planning for supplementary lessons of practical nursing class.
    - Students are freshmen and sophomores.
    - They participate in class voluntary for 3 months.

    b. Do

    - The chief instructor decides to put students into teams through her experience.
    - They are assigned to complete presentation of their projects with team members.

    c. Check

    - Evaluation of each team's assignment.
    - Reporting about each students' attitude or behavior in the team.

    d. Action

    - After they have finished their presentation, students take psychological testing.
    - The instructor will be able to predict which teams might succeed and which might not, referring to the results of her evaluation and observation of students' collaborative work in the team.
    - We asked her to compare her reflection with the evaluation of psychological testing so that she might become aware of the method of grouping.

2. **The second cycle** (a practical experiment)

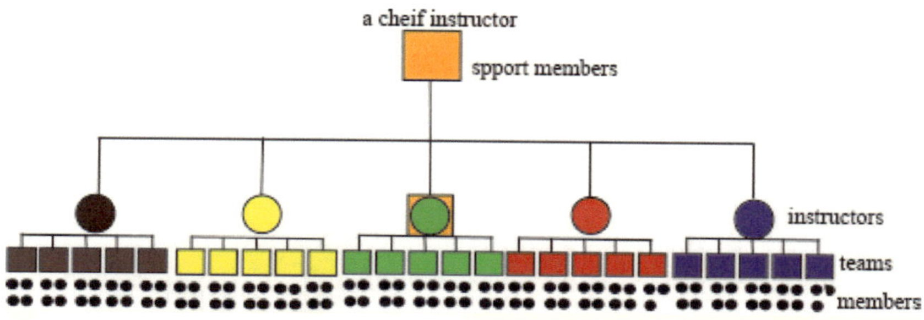

Figure 6.
The method of teaming [19].

a. Start: Data gathering for freshmen in 2015 (feedforward to Plan)

- Psychological testing (personality measurement).
- Questionnaire (ask about students' needs, requirements).
- Students' records in the previous year (in 2014).

b. Plan

- Asking the chief instructor to plan her first cycle reflection of the pilot experiment in order to decide members of teams [10] and form teams dividing them into five instructors (**Figure 6**).
- Making plan to collect data for evaluation of this support system.

c. *Do*: Explaining about the aims of PELS (feedforward to students) and gathering data (feedback to Check)

- Reports, attendance, and attitude (participant observation).
- High-stakes assessments.
- Examination of first term test for practical performances.
- Low-stakes assessments: pre- and post-quiz about basic nursing special concepts.

d. *Check:* Analysis and assessment of students' performance through data gathered in previous sections, Start, Plan, and Do

e. *Action*: Reflecting the classes by analyzing the result of students' performance and participant observation

5. Results

5.1 Pilot

At first, we have asked the chief instructor to reflect grouping of team members based on her experience, comparing with assessment of psychological testing and

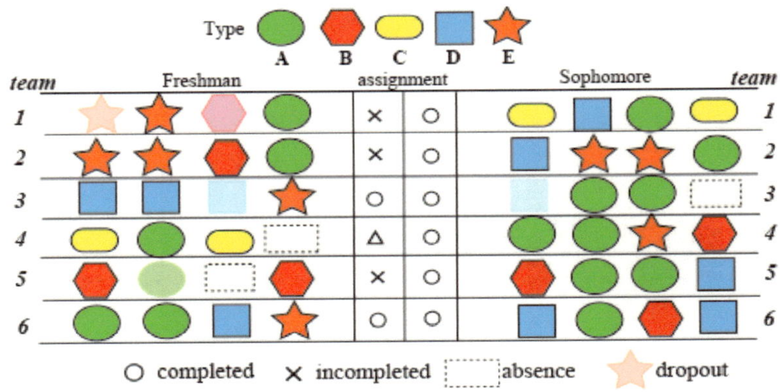

Figure 7.
The results of teams put together through the experience of instructors [19].

the other students' data for participants in the pilot experiment. We also have asked her to report them and explained her why some teams have not completed their presentation for their projects and the others have done well from the view points of matching members for each team.

The results of the pilot experiment are shown in **Figure 7**.

- Fifty-one students participated voluntarily (female; freshmen, 29; sophomore, 22).
- The period was from November in 2014 to January in 2015 (about 3 months).
- Fifteen groups were divided, not by assessment of psychological testing, but by instructor's experience (nine teams for freshmen, six teams for sophomores (**Figure 7**) [19].
- Two of nine teams have completed their presentation for freshmen, and four of nine teams dropped out. All teams of sophomores have completed and succeeded their presentation.

5.2 Examinations

We have done a practical examination for our PELS system along with the design of SPDCA cycle.

3. Practical experiments

   a. Participants: Ninety-eight freshmen students have participated in the practical examination.

   b. Period: The period was from April to August in 2015 (about 5 months).

   c. The data: All students' records were gathered (the first term test as high-stakes assessments and pre- and post-quiz as low-stakes assessment) the ratio of attendance and the record of the process how members have done their assignments interactively through text message, reports, and video which were recorded by the learning management system (LMS).

   d. Students' records of the previous year: the data of 97 freshmen's records in 2014 as high-stakes assessment.

e. Observation: Instructors report, and the participant observation was also gathered (**Table 1**).

5.3 Comparisons

1. The results of t-test probability (**Table 3**): There are comparisons of t-test results of probability between pre- and post-quiz for the low-stakes assessments, practical tests for the high-stakes assessments (**Table 2**), and the percentage of attendance between 2014 and 2015.

    a. Low-stakes assessment: The results of post-quiz about the special concept for nursing have become significantly higher than those of pre-quiz. The average point of the post-quiz was about 4.25 higher from 1.91 to 6.25 (total 10 points) (**Table 3**).

    b. High-stakes assessment: The result of practical assessments is significantly higher in 2015 than those in 2014. The average point was about 25 points higher from 58.94 to 83.95 (for a total of 100 points) (**Table 2**).

    c. Attendance: The result of the attendance percentage is higher in 2015 than that in 2014. The average of the potion of perfect attendance for the first semester in 2015 became higher from 97.3% in 2014 to 99.5%.

2. Low-stakes assessment (**Figure 10**): The concepts of nursing development: **Figure 8** shows the comparison of total points for each team between pre-and post-quiz in practical nursing class. The results of most teams were better in pro tests than pretest. The members of team 4 and 14 have declined their performance.

3. High-stakes assessment (**Figure 9**): The figure shows the distribution of students' performance records for practical assessment tests, comparing the results of the pre-term between 2014 and 2015. The range of 51–60 points is the highest in 2014 and those of 91–100 points in 2015. In the case of students whose performance are evaluated under 60 points, they might fail in the practical subject and drop out, except taking an examination again and acquiring credits.

Practical Assessment (100 points)							
Year	*N*	$\overline{X}$	*SD*	*df*	*F*	*test*	
2014	97	58.9	12.4	193	198.8***	0.000	***
2015	98	84.0	12.4				

Table 2.
Significant differences of high-stakes assessment.

Low Stakes Assessment		*N*	$\overline{X}$	*S.D.*	T-value	*df*	*t*- Test	
Quiz (10 points)	Pre	97	1.91	2.70	-11.6	96	0.000	***
	Post	97	6.25	3.19				

Table 3.
Significant differences of low-stakes assessment.

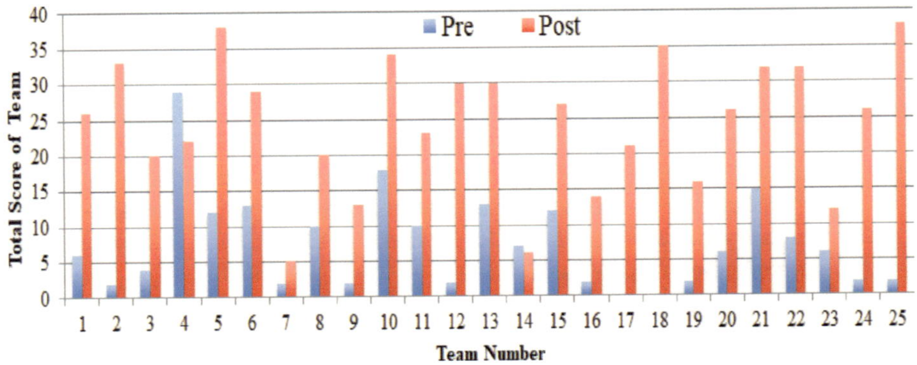

Figure 8.
Comparison of scores between pre- and post-quiz in practical experiment.

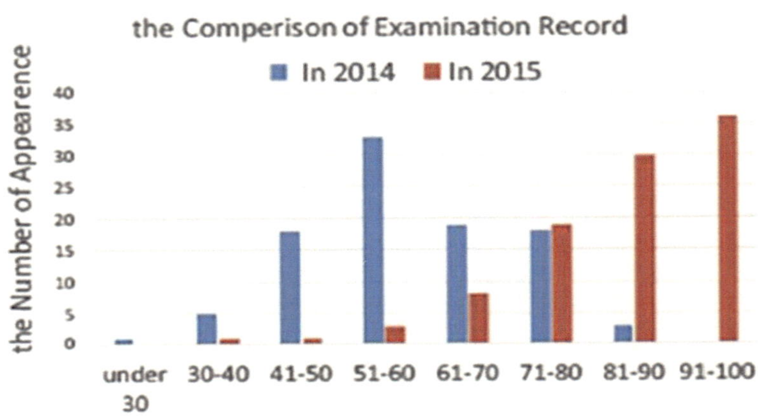

Figure 9.
Distribution of records comparing high-stakes assessment between 2014 and 2015.

4. The ratio of perfect attendance comparing between 2014 and 2015 (**Figure 10**): The figure shows that 96% of students had attended practical nursing classes perfectly in 2015, which is higher when compared to 53% in 2014.

5. Observation (instructors) (**Table 2**, (a)–(e)): The comparison of instructors' observation between 2014 and 2015: From their assessments of observation, students' performance has become better in 2015 than in 2014. The results are strongly related to other assessments;, for example, the situation of standby class is related to perfect attendance percentage.

6. Observation (instructors' reflection in class) (**Table 2**, (f)): Instructors have changed their views and attitude toward students, from subjective to objective.

7. Observation (supporters' participant observation in class) (**Table 2**, (g)): This is the supporters' reports on how the method of groping or teaming has been changed from 2014 to 2015. This method was consulted by supporters for instructors and students as the one element of the PELS system.

8. All results of observation reports from instructors (a)–(e) (**Table 1**) shows that the system has worked well with interactive communication among students which had been observed at role-playing in class (**Figure 11**).

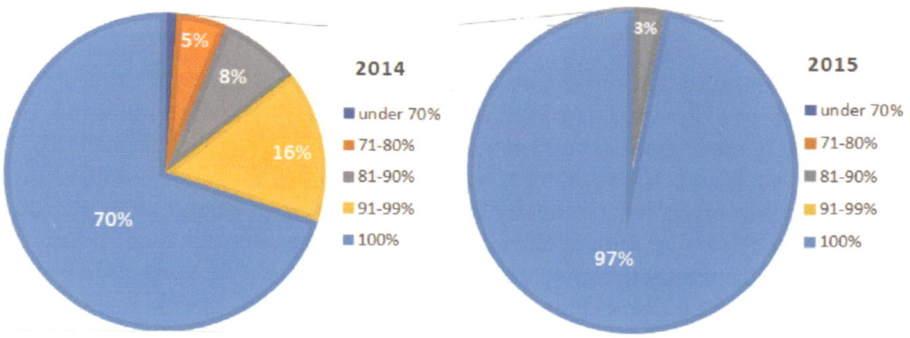

Figure 10.
Comparison of attendance ratio between 2014 and 2015 of the first semester.

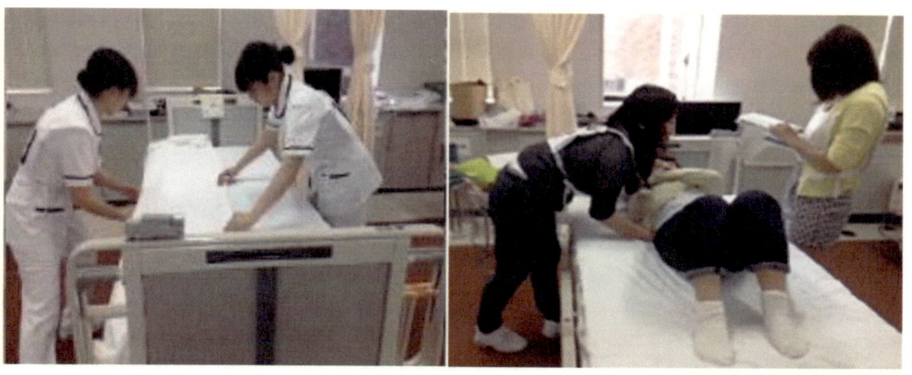

Figure 11.
Participant observation for role-playing in class [19].

6. Discussion

6.1 Examination of PELA

The results of examination for the PELS system were clear evidences that have shown fair improvement in students' performance and attitude. It might be able to say that the system has supported effectively for both instructors and students.

However, we will have to look at the reason why students have been hesitant in joining practices in class before the support had been conducted. Almost all students have begun to study about nursing enthusiastically and independently after the system has been provided. We might not be able to continue using this support system successfully or fruitfully, without finding out the reason why the system had an effect on nursing personalized education and learning [45, 46].

6.2 Examination of pre- and post-quiz

Pre- and posttests in class were the same tasks given to students (**Figure 8**, **Table 3**). Students were required to discuss the problems about the environment in a sickroom which was shown by a picture. This test measures how the students' concept has come to change, concerning their specialty in nursing. From the results, we have found that there has been a differentiation among teams, depending on students' traits and social abilities [47, 48]. Their interactive communication must have had an effect on their learning which might have changed their concepts for collaborative working. We will have to find out the reasons why the performance of some teams has significantly improved.

6.3 Observation

From this observation report, we can see that students had paid their big efforts to improve their performance in 2015 (**Table 2**). In the last column, instructors have reflected on their attitude toward students. We will need to consider about what is the reason for which instructors have been changed their view of teaching objectively and why they have become to change their attitude to students in class [49]. One of the reasons might have been the feedforward control system that enabled instructors to obtain students' data when they designed their instructions, depending on their personality, needs, and so on (feedforward control) (hypotheses (a)). They tried to analyze students' data in order to understand their differentiation. It means that the instructors began to see individual students and they might have obtained reliabilities from students (feedback control). I think this has an important meaning for personalized education and learning. We can say that it has brought them mutual understanding and interactive communication between not only instructors and students but also between students and students. Consequently, students might have shown their ability independently and positively to us (feedback control) (hypotheses (b)).

6.4 The relation to the theory of HAL

What is of most importance in the theory should be discussed. If students or instructors cannot find their own methods how they learn for changing their views, their improvement would not be continued (SPDCA cycle: **Figure 5** related to **Figures 1** and **4**). Instructor might have acquired how they can make an accurate forecast their students' behavior (Feedforward Control). According to it, they have planned their instruction for class and combined appropriate grouping and teaming. In practical class, it was considered that instructors have guided student adequately depending on their individual traits (Feedback Control A). Consequently, it might be observed interactive communication not only among students but also between instructors and students. Moreover, students' performance might have significantly improved with their autonomous and activity (Feedback Control B).

7. Conclusions

The purpose of this chapter is to propose the educational reform [3] so that we can maintain quality education. The collaborative learning with ICT might have been considered the best method to realize it, but there were some reported problems, especially regarding interactive communication (**Figure 2**) [13].

Along with the educational reform and personalized education and learning (PEL) [5], we have introduced cybernetics theory of HAL into our model of PELS, which consists two kinds of feedback control, feedforward control and main controller (**Figure 1**) [36].

Various data of learners' individual traits would be gathered and analyzed in the main controller (**Figure 3**). The results of those assessments would be informed to instructors before designing class (feedforward control) (**Figure 4**) [41]. Learners would be able to learn how to interact with others autonomously (feedback control B) (**Figure 1**).

The SPCDA cycle was applied (**Figure 5**) for experiments in nursing science class with problem solving so that we can prove whether the model of PELS is effective or not for collaborative learning. The result of the first cycle (pilot experiment) was shown in **Figure 7**. The instructors have obtained the rule for combination of team members (feedback control A) (**Figure 7**).

In the second cycle of SPDCK (practical experiment in 2015) they have decided how to combine members for optimized each team from students' traits which have been assessed by big data processing system (feedforward control) (**Figures 1** and **3**) [19]. The results of learning effects were shown by comparisons (**Tables 1** and **3**). Both low-stakes assessment (**Figure 8**) and high-stakes assessment (**Figure 9**) have been shown to be more significant. And moreover, the comparison of attendance ratio between 2014 and 2015 was shown to be more significant (**Figure 10**).

It is considered that the learning effects were shown significantly getting better by PEPS. It is considered the reports of participant observations, which were compared with students' attitude in class, matched with the results of assessments.

In conclusion, we have shown that the PELS works effectively with interactive communication between instructor and learner and learner and learner. Consequently, learners have become stronger to study independently and positively through personalized education and learning. This means that this system was useful and helpful for instructors by using big data processing and analysis [22] so that we can maintain quality education.

Acknowledgements

The author is grateful to Dr. Kiyoko Tokunaga and participants for collaboration on our practical research.

Author details

Keiko Tsujioka
Institute for Psychological Testing, Osaka, Japan

*Address all correspondence to: keiko_tsujioka@sinri.co.jp

References

[1] Miller R. Personalizing Education: Schooling for Tomorrow. Center for Educational Research and Innovation, OECD. Think Scenarios, Rethink Education, Schooling for Tomorrow. Paris: OECD Publishing; 2006. DOI: 10.1787/9789264023642-en

[2] Paludan JP. Personalized Learning 2025, in Personalizing Education, OECD/CERI. 2006. Available from: http://www.oecd.org/site/schoolingfortomorrowknowledgebase/themes/demand/41176429.pdf

[3] OECD. Personalising Education, Schooling for Tomorrow. Paris: OECD Publishing; 2006. Available from: https://doi.org/10.1787/9789264036604-en

[4] Curry L. An Organization of Learning Styles Theory and Constructs, Paper presented at the Annual Meeting of the American Educational Research Association1983. p. 28

[5] The Glossary of Education Reform (HP): Personalized Education. Available from: https://www.edglossary.org/personalized-learning/ [Accessed: 02 February 2020]

[6] Baylari A, Montazer GA. Design a personalized e-learning system based on item response theory and artificial neural network approach. Expert Systems with Applications. 2009;**36**(4):8013-8021

[7] Chena CM, Duh LJ. Personalized web-based tutoring system based on fuzzy item response theory. Expert Systems with Applications. 2008;**34**(4):2298-2315

[8] Kogo C, Tominaga A, Ishikawa N. Successful Implementation of University Blended Learning. Japan Society for Educational Technology. 2012;**36**(3):281-290

[9] Chen W, Looi CK, Looi CK. Incorporating online discussion in face to face classroom learning: A new blended learning approach. Australasian Journal of Educational Technology. 2007;**23**(3):307-326

[10] Osguthorpe RT, Graham CR. Blended Learning Environments: Definitions and Directions. The Quarterly Review of Distance Education. 2003;**4**(3):227

[11] Curtis JB, Charles RG, editors. The Handbook of Blended Learning; Global Perspectives, Local Designs. Michigan: John Wiley & Sons, Inc.; 2006

[12] Stahl G, Koschmann G, Southers DD. Computer supported collaborative learning. In: Sawyer RK, editor. The Cambridge Handbook of the Learning Sciences. Cambridge: Cambridge University Press; 2006

[13] Koschmann T. Paradigm shifts and instructional technology. In: Koschmann T, editor. CSCL: Theory and Practice of an Emerging Paradigm. Mahwah, NJ: Lawrence Erlbaum Associates, Inc.; 1996. pp. 1-23

[14] Fransen J, Weinberger A, Kirschner PA. Team Effectiveness and Team Development in CSCL. Educational Psychologist. 2013;**48**(1):9-24. DOI: 10.1080/00461520.2012.747947

[15] Kreijns K, Kirschner PA, Vermeulen M. Social Aspects of CSCL Environments: A Research Framework, Educational Psychologist. Educational Psychologist. 2013;**48**(4):229-242. DOI: 10.1080/00461520.2012.750225

[16] Guetzkow H, Gyr J. An analysis of conflict in decision making groups. Human Relations. 1954;**7**:367-381

[17] Katz N, Lazer D, Arrow H, Contracto N. Network theory and small groups. Small Group Research. 2004:307-332. DOI: 10.1177/1046496404264941

[18] Cronbach LJ, Snow RE. Aptitudes and Instructional Methods. Irvington, New York: John Wiley & Sons Inc.; 1997

[19] Tsujioka K. A case study of using big data processing in education: The method of matching members by optimizing collaborative learning environment. In: Social Media and Machine Learning. IntechOpen; 2019. DOI: 10.5772/intechopen.85526

[20] Baker R, Simens G. Educational data minding and learning analytics. In: Sawyer RK, editor. The Cambridge Handbook of the Learning Sciences Second edition. Cambridge: Cambridge University Press; 2014. pp. 253-271

[21] Baker RS. Minding data for student models. In: Nkambou R, Bourdeau J, Mizoguchi R, editors. Advances in Intelligent Tutoring Systems. Studies in Computational Intelligence. Vol. 308. Berlin, Heidelberg: Springer; 2010. pp. 323-337

[22] Tsujioka K. A Case Study of ICT Used by Big Data Processing in Education: Discuss on Visualization of RE Research Paper. ICIET, Association for Computing Machinery; 2018

[23] Tsujioka K. Toward clarifying human information processing by analyzing big data: Making criteria for individual traits in digital society. In: Beatrice O, editor. Strategy and Behaviors in the Digital Economy. London: IntechOpen; 2019. DOI: 10.5772/intechopen.86037

[24] Márquez-Vera C, Cano A, Romero C, Noaman AYM, Fardoun HM, Ventura S. Early dropout prediction using data mining: a case study with high school students. Expert Systems. 2016;**3**(1): 107-124. DOI: 10.1111/exsy.12135

[25] Márquez-Vera C, Cano A, Romero C, Ventura S. Predicting student failure at school using genetic programming and different data mining approaches with high dimensional and imbalanced data. Applied Intelligence. 2013;**38**(3):315-330. DOI: 10.1007/s10489-012-0374-8

[26] Hu H, Wen Y, Chua T, Li T. Toward scalable systems for big data analytics: A technology tutorial. IEEE. 2014;**2**:652-687. DOI: 10.1109/ACCESS.2014.2332453

[27] Jonassen DJ, Peck KI, Wilson BG. Learning with Technology: A Constructivist Perspective. Prentice Hall; 1999

[28] Pietrazak M, Palisziewiz J. Framework of strategic learning: The PDCA cycle. Management. 2015;**10**(2):149-161

[29] Matsuo M, Nakahara J. The effects of the PDCA cycle and OJT on workplace learning. The International Journal of Human Resource Management. 2013;**24**(1):195-207. DOI: 10.1080/09585192.2012.674961

[30] Tsukahara A, Hasegawa Y, Eguchi K, Sankai Y. Restoration of Gait for Spinal Cord Injury Patients Using HAL with Intention Estimator for Preferable Swing Speed. IEEE Transactions on Neural Systems and Rehabilitation Engineering. 2015;**23**(2):308-318. DOI: 10.1109/TNSRE.2014.2364618

[31] Sakurai T, Sankai Y. Development of Motion Instruction System with Interactive Robot Suit HAL, International Conference on Robbotics and Biomimetics2009. pp. 19-23

[32] Kawamoto H, Sankai Y. Comfortable power assist control method for walking aid by HAL-3. In: IEEE International

Conference on Systems, Man and Cybernetics, Yasmine Hammamet, Tunisia. Vol. 4. 2002. pp. 6-12. DOI: 10.1109/ICSMC.2002.1173328

[33] Sakurai T, Sankai T. Development of motion instruction system with interactive robot suit HAL. In: IEEE International Conference on Robotics and Biomimetics (ROBIO), Guilin. 2009. pp. 1141-1147. DOI: 10.1109/ROBIO.2009.5420755

[34] Suzuki K, Mito G, Kawamoto H, Hasegawa Y, Sankai Y. Intention-based walking support for paraplegia patients with Robot Suit HAL. Advanced Robotics. 2007;**21**(12):1441-1469

[35] Sankai Y. Leading edge of cybernics: Robot suit HAL. In: SICE-ICASE International Joint Conference, Busan. 2006. pp. 1-2. DOI: 10.1109/SICE.2006.314982

[36] Tsujioka K. Development of support system modeled on robot suit HAL for personalized education and learning. In: 2017 International Conference of Educational Innovation through Technology (EITT). Osaka: IEEE; 2017. pp. 337-338

[37] Kawamoto H, Sankai Y. Power assist system HAL-3 for gait disorder person. In: Miesenberger K, Klaus J, Zagler W, editors. Computers Helping People with Special Needs. ICCHP 2002. Lecture Notes in Computer Science. Vol. 2398. Berlin, Heidelberg: Springer; 2002

[38] Tsukahara A, Kawanishi R, Hasegawa Y, Sankai Y. Sit-to-stand and stand-to-sit transfer support for complete paraplegic patients with robot suit HAL. Advanced Robotics. 2010;**24**(11):1615-1638. DOI: 10.1163/016918610X512622

[39] Kawamoto H, Sankai Y. Power assist method based on phase sequence and muscle force condition for HAL. Advanced Robotics. 2005;**19**(7):717-734. DOI: 10.1163/1568553054455103

[40] Kawamoto H, Kadone H, Sakurai T, Sankai Y. Modification of hemiplegic compensatory gait pattern by symmetry-based motion controller of HAL. In: 37th Annual International Conference of the IEEE Engineering in Medicine and Biology Society (EMBC). Milan; 2015. pp. 4803-4807. DOI: 10.1109/EMBC.2015.7319468

[41] Hayashi T, Kawamoto H, Sankai Y. Control method of robot suit HAL working as operator's muscle using biological and dynamical information. In: IEEE/RSJ International Conference on Intelligent Robots and Systems. Edmonton: Alta; 2005. pp. 3063-3068. DOI: 10.1109/IROS.2005.1545505

[42] Kasaoka K, Sankai Y. Predictive Control Estimating Operator's Intention for Stepping-up Motion by Exo-Sckeleton Type Power Assist System HAL, IROS20012001. pp. 1578-1583

[43] Sankai Y. Leading edge of cybernics: Robot suit HAL. In: SICE-ICASE International Joint Conference. Busan; 2006. pp. P1-P2. DOI: 10.1109/SICE.2006.314982

[44] Dannelle DS, Antonia JL. Introduction to Rubrics: An assessment tool to save grading time. Convey Effective Feedback, and Promote Student Learning. Virginia: Stylus Publishing; 2013

[45] Munakata M, Vaidya A. Fostering creativity through personalized education. Primus. 2013;**23**(9):764-775. DOI: 10.1080/10511970.2012.740770

[46] Waldeck JH. Answering the question: Student perceptions of personalized education and the construct's relationship to learning outcomes. Communication Education. 2007;**56**(4):409-432. DOI: 10.1080/03634520701400090

[47] Guilford JP. Fundamental Statistics in Psychology and Education. N.Y.: McGraw-Hill; 1965

[48] Tsujioka B, Sonohara T, Yatabe TA. Factorial study of the temperament of Japanese College Male students by the Yatabe-Guilford Personality. Psychologia. 1957;**3**:110-119

[49] Dumitru E, Pierre-Antoine M, Bengio Y, Samy B, Pascal V. The difficulty of training deep architectures and the effect of unsupervised pre-training. Journal of Machine Learning Research. 2009;**5**:153-160

Chapter 7

Developing a Cloud Computing Framework for University Libraries

Samuel Owusu-Ansah, Stephen Budu and Rosina Akuokor Anati Budu

Abstract

Our understanding of the library context on security challenges on storing research output on the cloud is inadequate and incomplete. Existing research has mostly focused on profit-oriented organizations. To address the limitation within the university environment, the paper unravels the data/information security concerns of cloud storage services within the university libraries. On the score of changes occurring in the libraries, this paper serves to inform users and library managers of the traditional approaches that have not guaranteed the security of research output. The paper is built upon the work of Shaw and the cloud storage security framework, which links aspects of cloud security and helps explain reasons for university libraries moving research output into cloud infrastructure, and how the cloud service is more secured. Specifically, this paper examined the existing storage carriers/media for storing research output and the associated risks with cloud storage services for university libraries. The paper partly fills this gap by a case study examination of two (2) African countries' (Ghana and Uganda) reports on research output and cloud storage security in university libraries. The paper argues that in storing university research output on the cloud, libraries consider the security of content, the resilience of librarians, determining access levels and enterprise cloud storage platforms. The interview instrument is used to collect qualitative data from librarians and the thematic content analysis is used to analyze the research data. Significantly, results show that copyright law infringement, unauthorized data accessibility, policy issues, insecurity of content, cost and no interoperable cloud standards were major risks associated with cloud storage services. It is expected that university libraries pay more attention to the security/confidentiality of content, the resilience of librarians, determining access levels and enterprise cloud storage platforms to enhance cloud security of research output. The paper contributes to the field of knowledge by developing a framework that supports an approach to understand security in cloud storage. It also enables actors in the library profession to understand the makeup and measures of security issues in cloud storage. By presenting empirical evidence, it is clear that university libraries have migrated research output into cloud infrastructure as an alternative for continued storage, maintenance and access of information.

Keywords: cloud computing, storage, security, research output, academic libraries, university libraries

1. Introduction

Despite the renewed interest in safeguarding research output, the changing storage carriers due to the fragility of storage carriers, lifespan, and handling practices are a cause of concern for the university libraries [1]. University libraries cannot avoid working in the cloud as they have become adaptive to inevitable and unpredictable changes occurring within the digital environment [1]. The university community places much emphasis on research and publication not only because it is presumed that research enriches teaching and the learning process, contributing to the body of knowledge, but also because it is a major determinant of institutional prestige and that of the nation at large [2].

Irrespective of the technological changes, stored research output in universities must be secured for future availability and accessibility [3]. Cloud storage has become an alternative for the storage of research output. According to Yuvaraj [4], university libraries have continued not only as only new technology adopters but rather cutting-edge IT users. Clearly, cloud computing as a cutting-edge IT platform proves to be a lasting technological innovation that continues to rise in usage [5].

However, owing to the technological age, university libraries are faced with new opportunities for innovative educational practices, hence providing electronic library services. Almost all university libraries are primarily concerned with enhancing teaching, learning, and research through the provision of timely information resources. On that basis, researches by Gabridge [6], Gold [7] and Jones [8] revealed the need for libraries to provide research data services. In providing timely information resources, modern libraries' digital collections must be stored for future use and as backups to ensure continuous accessibility by library users.

Witten and Bainbridge [9] explained that a digital library is a focused "collection of various forms of digital objects" such as text, audio, and video, as well as their methods for access, retrieval, selection, organization, and maintenance. Rosenberg [10] also reiterated that a digital library can refer to information resources which are accessed by and delivered to users electronically or via a network [11]. Primarily, in developing countries, microfilms, databases, CD-ROMS, hard disks, external drives have been the existing platforms for storing library digital information, though these come with major drawbacks. For instance, these storage devices are exposed to threats such as theft, inadequate storage space, virus attacks and unauthorized accessibility among others. These drawbacks have been a major concern for academic libraries' thus an ongoing debate and discussion on the new technology "cloud storage" as an alternative storage media.

To a large extent, studies confirm that modern university libraries have greatly shifted from traditional roles (paper-based services) to digital library services. This paradigm shift has paved the way for library services to be accessed and delivered via the web [12]. For university libraries, the issue of using cloud services to store digital collections is particularly important as technological changes have paved the way for library services to be accessed and delivered via the web [12]. As more data and information is generated and stored in the cloud, either by design or default, university libraries need to be confident of the security of the digital collections. There is a growing interest in the implementation of cloud storage services which exposes university libraries to a new set of threats and vulnerabilities. McLeod and Gormly [13] concluded that if cloud service providers are to be used, their security, viability, sustainability, and trustworthiness must be paramount.

Studies have demonstrated that that cloud computing in libraries has widely examined the rise of data-intensive services in academic libraries with less emphasis on cloud storage security [14, 15]. Most of these studies were based on individual or small-scale survey data concentrated in one country. Owing to the extant gap in wide-scale exploratory studies, the present paper explored the risks associated with cloud storage services and how university libraries can ensure safe research output. In this light, the paper contributes significantly to the body of literature by unraveling new evidence from universities located in Ghana and Uganda on how academic libraries can secure research output with cloud services.

The following sections include research questions, related literature, theoretical framework, research methodology, results, a summary of key findings, conclusion and recommendation.

1.1 Research questions

i. What are the existing storage carriers/media for storing research output in university libraries?

ii. What are the reasons for university libraries moving research output into a cloud infrastructure?

iii. What risks are associated with cloud storage services for university libraries?

iv. In university libraries, how can research output store on the cloud service be secured?

2. Related literature

2.1 Storage media in university libraries

Libraries use several types of media in storing digitized content or information (audio, video, text, images etc.). Each of the media suffers disadvantages with regard to reliability, high lifespan, ease of access and validation plus various costs. Enakrire and Baro [16] argued that these media include;

i. **Magnetic disk drives** are disk drives which are mostly mounted on computers. They are inexpensive, of very high-density, fast to use, and multiple user connectivities to the server are possible.

ii. **Magnetic tape**, which comes in various formats and can only be effective for duplicate or backup copies. However, they are not recommended for primary storage.

iii. **Optical disks**, for example, CD-R and DVD-R cost less, use low energy but exert high labor costs, poor accessibility, a periodic verification is not cost-effective and low density by today's standards. Others are CD-RW and DVD-RW these are recommended for individual and day-to-day use but are not recommended for data preservation [17].

2.2 Cloud storage services

Until recently, evidence from the pool of literature shows that the concept of cloud is of the growing research area. Indeed, a lot more storage capabilities exist in the cloud. According to Mavodza [18], cloud computing is the delivering of hosted electronic services over the internet. Scale [19], opines that it is: "the sharing and use of applications and resources of a network environment to get work done without concerns about ownership and management of the network's resources and applications, data are no longer stored on one's personal computer, but are hosted elsewhere to be made accessible in any location and at any time". Gosavi et al. [20] iterated that cloud computing harnesses the capabilities of resources like storage, scalability, and availability, which are accessible to university libraries as clients. Hence, depending on the needs of the clients, the infrastructure can be scaled up or down.

In developed or developing countries, cloud storage provides promising advantages to university libraries. According to Li [21], cloud storage reduces the cost of hardware and software, and it makes the storage and management of data on the internet possible. It also reduces the work of Information Technology (IT) professionals as most of the system's work is performed by the hosting company. Payment for the cloud storage service is by pay-as-you-go, which is convenient for organizations such as academic libraries which have budget restraints. Han [22] enumerates cost-effectiveness, flexibility, and data safety as a rationale for cloud storage in academic libraries. Han [23] alludes the advantages that cloud storage has over traditional storage to "availability, scalability, off-site storage, on-demand, and multi-tenancy" which allows different applications or different users to access the same resources to fit their needs. Han further states that data stored in the cloud can be easily transferred and duplicated globally to minimize data loss due to natural disasters.

Haris [24] also gives an analysis of the benefits of cloud storage especially for libraries and these include high performance, an avenue for collaboration, less "need for in-house technical expertise, cost savings, and more timely access" to the latest IT functionality. Haris further states that the cloud also provides a better workflow, "automated software updates, redundancy", and backups. Cloud storage provides collaboration, particularly for academic and research libraries. Through the use of cloud technology, a collaboration between libraries, researchers, and students is promoted. The cloud also enables remote access to a wide range of research materials.

2.3 Reasons for cloud computing in university libraries

In this section, the role of cloud computing in university libraries, specific cloud storage platforms and the risks associated with cloud storage are reviewed.

Kaushik and Kumar [25] contend that cloud computing can offer many interesting possibilities for institutions such as libraries. Cloud computing is quite significant as it reduces technology cost, increases capacity reliability, and storage performance for some type of automation activities like library services. In recent times, cloud computing has made strong inroads into other commercial sectors and is now beginning to find more of its applications in the library and information environment.

After the personal computer and the internet, cloud computing also known as the third revolution is completely new in terms of technology. Potentially, cloud computing is an unraveled technology in university libraries as digital content can be stored in the cloud. Mobile devices are enabled using cloud computing by taking out an item or scanning a barcode [26]. Gosavi et al. [20] argued that when using cloud computing, users can be able to browse a physical shelf of books located in the library, choose an item or scan a barcode into his mobile device. More so, heritage materials or documents can be digitized, searched and accessed by library patrons. The new concept of cloud libraries includes OCLC, Library of Congress (LC),

Exlibris, Polaris, Scribd, Discovery Service, Google Docs/Google Scholar, WorldCat and Encore [27].

Nowadays, studies appear to be emerging in cloud computing. For instance, a paper presented by Saleem et al. [27] indicated that university libraries have adopted cloud computing technology to enhance library services by adding more values, attracting users and cost-effectiveness. In the cloud computing environment, clouds have vast resource pools with on-demand resource allocation and a collection of networked features. The new concept of cloud and libraries has generated a new model called cloud libraries.

In the work of Zainab et al. [28], it was reported that the first reason of shifting research report into cloud computing is to reduce the total cost of ownership and maintenance of the cloud infrastructure. Secondly, scalability of the cloud service system is another objective, so that it is able to handle increased traffic. Due to the rapid expansion of the user group, we need to redesign the back-end web server with scalability in mind, such that it is able to accommodate an increasing number of concurrent users.

Based on the web traffic statistic, the average visit per month for the year 2012 is approximately 87,000 users and we expect the numbers will grow in the coming years as resources in the repository also grew. The high volume of transaction is causing The server to behave extremely sluggish and crashes frequently [28]. On the hand, migration is necessary in order to meet the increasing demand for storage space for full-text digital resources. File sizes of some digital resources are extremely large especially audio, video and images. Besides, as more users' access and upload articles to the magnetic hard drives, university libraries face problems in fulfilling the storage space demand. The cloud storage service which promises and contributes to about 13 terabytes of storage space, can store over 12 million digital files of research output. Thus, it is very obvious that without a long-term plan, university libraries would not able to sustain the present storage demand from users in the future until alternative storage is assessed.

It is expected that migration of digital files would reduce downtime when scheduled backup and indexing, as well as site traffic, occur simultaneously. The previous system backup was very laborious and time-consuming. Often scheduled jobs would cause unnecessary downtime of the magnetic and optical systems. System downtime is unavoidable because the system was hosted without a redundant server.

2.4 Enterprise cloud storage platforms

Amazon S3: Amazon Simple Storage Services (Amazon S3) provides a secure, durable, highly-scalable object storage (Amazon, 2015). It uses a web service interface to store and retrieve any amount of data. It is a pay as you use service. There are different storage classes designed for different uses; Amazon S3 standard, Glacier for long-term archive. The services include backup and archiving, disaster recovery, and big data analytics [29].

Google cloud storage: Allows storage and retrieval of any amount of data at any time. It facilitates the storage of data on Google's infrastructure with high-reliability performance and availability (Google, 2015) [29]. The services include data storage, large unstructured data objects, uploading data, and managing data. The lowest storage class is $0.01 GB/month.

Microsoft Azure: Azure supports the selection of wide services including operating systems, frameworks, tools, and databases. It's typically a platform-as-a-service and software-as-a-service. It provides secure private connections, storage solutions, and data residency and encryption features (Microsoft, 2015). It provides scale-as-you-need, pay-as-you-go service plan, and strong data protection security.

Other cloud storage platforms include Dropbox, SkyDrive, Box, Google Drive, Flickr, Google music, Apple iCloud, and Amazon cloud player.

2.5 Associated risks and possible solutions for cloud storage

Lili and Buer [30], highlighted that advancement in technology may not necessarily transform the cloud services into mainstream technology in academic libraries. A scan of literature [31–33], revealed that cloud security, interoperability, and regulatory perspectives are worrying. In addition, academic libraries may or may not completely lose control over IT and data. Sometimes, trust in the service provider, data portability, migration, copyright issues, and privacy is a big risk when it comes to adopting cloud computing technology.

2.5.1 Policy issues

Policies guide institutions and operations on what to do and not to do. Cloud storage and applications are valuable resources that allow academic libraries to store large amounts of information and perform collaborative tasks more effectively. However, there are risks associated and that must be mitigated in order to properly secure the research assets placed into the cloud [32]. In this light, it is purposeful for the policy to provide the framework within which the libraries will be expected to operate for storage and process information in cloud environments. Basically, the policy should encompass the scope of work, software, research information, human resource, users, copyright and many more.

2.5.2 Unauthorized accessibility

Once a digital collection (scholarly works, publications/collections, and historical documents) is put on the cloud, it becomes available for all groups of users and this can be exposed to unauthorized access to data centers. "*Cloud operators can dictate the manner in which users can access, use and reuse content or information via specific online services or applications. That is, the user interface ultimately dictates what can or cannot be done by end-users, regardless of what they are theoretically entitled to under the law*" [34]. So, the question is whether academic libraries can allow such law to be overridden on as it has already fallen in the public domain. This indeed is likely to impact on copyright law in the context of online applications.

2.5.3 No interoperable cloud standards

Cloud storage service providers are not guided by standard regulations. As a result, some service providers are tempted to offer low-quality services to developing countries in Africa thus creating loopholes for cybercriminals to take advantage. As an emerging trend, this issue of no interoperability is of concern, if research assets can be secured on the cloud. Interoperability refers to the ability of a collection of communicating entities to share specific information and operate on it according to agreed-on operational semantics [35]. Even though the clients (academic libraries) desire standards for cloud interoperability, the reality currently is that standard efforts only focus on portability, which is the ability to migrate workloads and data from one provider to another.

Librarians cannot sit unconcerned in this matter since the open access (OA) repositories are also part of collections of the library [36]. Though the OA repositories facilitate sharing of resources in educational research through portals that are modeled as gates to several repositories, it is a challenge because data synchronization is an issue when components in different clouds or internal resources work together, whether or not they are identical. Communication between clouds typically has a high latency, which makes synchronization difficult. Also, the two clouds may have different access control regimes, complicating the task of moving data between them [37].

Thus, interoperability is required, not just between different components, but between identical components running in different clouds [38]. Such components often keep copies of the same data, and these copies must be maintained in a consistent state. The design approach must address management of "system of record" sources, management of data at rest and data in transit across domains that may be under control of a cloud service consumer or provider and data visibility and transparency.

Nurnberg et al. [39] argued that full interoperability includes dynamic discovery and composition: the ability to discover instances of application components and combine them with other application component instances, at runtime. Application interoperability requires more than communications protocols. It requires that interoperating applications share common processes and data models. These are not appropriate subjects for generic standards, although there are specific standards for some particular applications and business areas.

2.6 Cost

Obviously, the cost is a challenge for academic libraries. More especially, enterprise cloud storage platforms such as Amazon S3 and Microsoft Azure are paid for as you use the cloud services. Unfortunately, libraries that find it difficult to fund basic services will see that as an extra cost inhibiting them to withdraw from the cloud service. The cost comes with human resource and sometimes maintenance of servers.

3. Theoretical framework for cloud storage security

The paper adopts the development of a Cloud Storage Security Framework (CSSF) to support an integrative approach to understanding and evaluating security in cloud storage in university libraries. The framework enables understanding

Factor	Item
1. **Cloud Storage Security**	1. Cloud Storage Security Policies
	2. Cloud Storage Security Procedures
2. **Confidentiality**	3. Identification of cloud storage user
	4. Authorisation to access data
3. **Integrity**	5. Accurate ownership of data
	6. Encryption of data
4. **Availability**	7. Accessible to the data
	8. Up-to-date available data
5. **Non-repudiation**	9. Accurate time-stamping of accessed data
	10. Assurance with user signature
6. **Authenticity**	11. Verified data based on authentication
	12. Synchronised data in the storage
7. **Reliability**	13. Consistency of cloud service
	14. Valid service
Factors: 7	Items: 14

Figure 1.
CSSF. Source: Yahya [40].

of the makeup of cloud storage security and its associated measures. Drawing upon CSSF, it indicates that security in cloud storage can be determined by seven factors: (1) security policies implementation in cloud storage, security measure that relates to (2) protecting the data accessed in cloud storage; (3) modifications of data stored; (4) accessibility of data stored in cloud storage; (5) non-repudiation to the data stored; (6) authenticity of the original data; (7) reliability of the cloud storage services.

The framework is summarized in **Figure 1**.

In applying the framework to the current research, security of research output in the cloud infrastructure can be determined by ensuring that all the seven factors are met by the university library.

4. Research methodology

This study aimed to explore security issues considered in migrating research output to the cloud service as input into the development of preservation or storage systems within the library environment. This section described an approach followed in the study. This included the research approach, purpose, instrumentation, and sources of data. Our paper adopted the qualitative approach to explore cloud computing in university libraries in the sub-Saharan Africa. Using a wide range of evidence and discovering new issues, the purpose of the paper was to explore the risks associated with cloud storage and security implications. The exploratory design was significant as the authors became more familiar with basic facts, settings, concerns, and generating new ideas. In this study, interviews were conducted with respective librarians in charge of research output within the (4) universities. Hence, the research sites were purposefully selected to ensure that they provided sufficient opportunities to test available infrastructure for storing research output. Again, since the paper was interested in only libraries with repositories, the institutions without OA repositories were excluded.

An interview schedule on the research questions was presented to 4 librarians from the universities. Thus, participants for the investigation were made up of librarians in charge of institutional repositories. These four university libraries selected were; Balme Library, (University of Ghana—Legon), Kwame Nkrumah University of Science and Technology—KNUST library (Ghana), HamuMukasa Library, (Uganda Christian University), and The Iddi Basajjabalaba Memorial Library, (Kampala International University—Uganda). The thematic content analysis was used to analyze the qualitative data. The authors further reviewed scholarly research articles, explored in the context of research data storage in and outside Africa.

5. Results

This section draws reference from respective university libraries in the context of cloud storage security for research data.

5.1 Balme Library, University of Ghana, Legon

The University of Ghana (UG), the premier university and the largest university in Ghana was founded as the University College of the Gold Coast by Ordinance on August 11, 1948, for the purpose of providing and promoting university education, learning and research. The vision of the university is "to become a world-class research-intensive University over the next decade". To achieve the vision, it "will create an enabling environment that makes the University of Ghana increasingly relevant to national and global development through cutting-edge research as well as high quality teaching and learning" (http://ug.edu.gh). To achieve this mountainous vision in the next decade, the Balme Library, the central library of the university plays a crucial role.

Established in 1948, the Balme Library is the main library of the University of Ghana. In addition to the Balme Library, there are other libraries in the various Schools, Institutes, Departments, Halls of Residence and the Accra City Campus which form the University of Ghana Library System (http://balme.ug.edu.gh).

In UG, research assets (theses, journals, newspapers) in the form of PDFs, word files, conference papers, videos, and audio have been generated. In the context of this study, the existing storage media for storing research data include CDs, DVDs, external drives, servers, hard drives, microfiche, and microfilms. Others include networked drives, Google drive and Dropbox used by researchers and the library in storing research assets.

The interviewee indicated that digital storage and backup is important because;

> *"Data may need to be accessed in the future to explain or augment subsequent research. Other researchers might wish to evaluate or use the results of previous research outputs as precedence to conduct other similar or extended studies".*

Agrawal and Nyamful [41] corroborated the findings in the present study. Accordingly, they reported that storage devices which stores and maintains large sets of data over time play an important role in mitigating big data challenges. Factors such as capacity, reliability, performance, throughput, cost, and scalability are involved in any ideal storage solution system. They argued that reliability is basically the retrieval of data in its original form without any loss. The issue of reliability takes into account both internal and external system failures and vulnerabilities. With the scale of data, the probability of losing some data during retrieval can be very high. In order to ensure continuous accessibility of data, storage is very necessary.

It was revealed by the interviewee that

> *"there is no robust or enough backup plan when the primary server goes down. With an average of 3000 visits per day on the Institutional Repository (IR), we wish to keep The website availability as high as possible. To solve the problem, the IR team decided to move digital files to a cloud environment using virtualization technology".*

A study by Ji et al. [42], revealed a compelling need for storage and management of research output. Given the current development of data (text, audio, video, images, etc.), university libraries are employing techniques such as data compression, deduplication, object storage, and cloud storage.

The Librarian in charge of research data opined that

> *"Unauthorized accessibility, physical damage, theft, and hacking are particular concerns with electronic data. Many research projects involve the collection and maintenance of human subject's data and other confidential records that could become the target of hackers and thus integrity must be maintained. The costs of reproducing, restoring, or replacing stolen data and the length of recovery time in the event of a theft highlight the need for protecting the computer system and the integrity of the data".*

The Librarian iterated that several issues are associated with storing research data on the cloud.

One interviewee pointed out that;

> *"Risks associated with cloud storage are crucial for the Balme Library. Storing research assets online via the Dropbox, mozy.com, Box.net, Adrive.com, Carbonite.com have proven the best alternative. However, a few associated risks include issues regarding property rights, copyright, data protection licenses or privacy. Other issues to consider is the fact that in the event of restoring data, it may be a bit slow and the service provider (Google Reader) could go out of service".*

5.2 KNUST

KNUST Library has realized the need to digitize and store documents and research data generated by staff and students of the University, hence the decision to create the online Institutional Repository. The online repository showcases the intellectual output from the KNUST. In the earlier 2010, a server and scanners were acquired to support digitization processes. Since then, postgraduate thesis, reports, and few research articles have been uploaded unto the repository. Increasingly, the project has continued to receive acclamation internationally due to robust IT infrastructure in the library.

The librarian for KNUST responded in this manner,

> *"Currently, the KNUST uses non-web based storage media to store data. There are two servers; one for the Library's catalogue and another for the Institutional Repository. The library also uses an external hard drive as a backup, but both media are located in-house".*

Reed et al. [43] asserted that "data backup plays an indispensable role in the computing system. Backup is one way to ensure data protection. By keeping copies of production data, backup protects data from a potential loss such as hardware and software loss, human errors, and natural disasters. The huge amount of data needing backup and archiving has reached several petabytes and may soon reach tens, or even hundreds of petabytes. The massive amount of data in today's library environment may consume much storage."

Furthermore, it was reported by the interviewee that

> *"The challenge faced with this kind of storage media is frequent memory crash, lack of expertise to manage the storage media, lack of space – the servers have low memory space, an interrupted power supply which uninterruptible power supply (UPS) is not even able to solve. Then finally, remote access to the information is denied because data is not online". Thus, the need to seek cloud storage.*

It was evident from the interviewees that cloud computing environments are easily scalable and backup recovery is very easy in Infrastructure as a Service (IaaS) Providers, hence there is efficient incident response whenever data needs to be recovered.

The authors sought to find what risks were associated with cloud storage. Cost and data security were concerns raised by the managers of the repositories. Agrawal and Nyamful [41] argued that the state of preventing a system from vulnerable attacks is considered as the system's security. Security risks involved with the use of cloud computing have various risk factors for the library environment. Seven important identity factors for risk in a cloud computing model are access, availability, network load, integrity, data security, data location, and data segregation.

5.3 Hamu Mukasa Library, Uganda Christian University Library

Uganda Christian University has been in existence for 11 years having only one library which uses traditional devices. In the year 2015, the library launched its institutional repository. The storage media for storing research data in Uganda Christian University library is examined as follows:

Uganda Christian University has both traditional and modern storage devices. Traditional storage includes CDs, flash disks, card catalog and later introduced modern storage like creating an institutional repository where dissertations and research papers are kept safely for future use.

The Librarian in charge of the research data output of the Uganda Christian University observed that;

> *"For modern storage devices, Google drive is currently used to store documents such as student Theses works, proposals, and the day to day statistics. This started early last year when the learning commons was opened. This is used because it is cheap and can be accessed easily by staff and students while doing their work".*

In this twenty-first century, information is not just in print but digitally created and reused by researchers and patrons within academic institutions. There is a need for digital information storage at Uganda Christian University because of the advantages. Prior to cloud storage, institutions invested heavily in data centers and servers even though they may not have used its storage space. The cloud storage allows institutions' (academic libraries) only pay for computing resources they use. By using cloud storage one can achieve a lower variable cost than can be gotten on the traditional storage devices.

However, using cloud storage by Uganda Christian University academic library has some risks. Lack of internet access or less bandwidth is a major issue. Specifically, when the internet is down its difficult for data to be retrieved thus inconveniencing the patrons. Secondly, sensitive information for the institution can be disclosed accidentally or deliberately in cloud services if not handled well especially when demand grows. Thus, the inappropriate accessibility of the institution data can be compromised.

For an institution like Uganda Christian University Library to ensure the safety of its research information in the cloud, the following must be considered.

- avoid unauthorized accessibility of research data using strong passwords.

- Privacy policy services settings must always be checked by appropriate management.

5.4 The Iddi Basajjabalaba Memorial Library, Kampala International University-Uganda

The Iddi Basajjabalaba Memorial Library (The IBML) is an integral part of Kampala International University (KIU). It is the intellectual hub of the university that supports the study, teaching, research and social information needs of the university. The IBML has grown over the years from one small room in 2001 manned by one member of staff and serving 700 users to an eight ultra-modern building serving over 20,000 users. The IBML system has evolved over time from the manual system providing print information resources to automated circulation services and digital information resources. In 2014, The IBML set up a digital repository to capture, store, and disseminate the intellectual content of the university. The digital content includes research articles, papers written by university staff, PhD theses, and other university publications. DSpace software was used for this project and it is hosted locally on a networked server. The repository data is backed up on an external hard drive with several terabytes of storage capacity.

The IBML has not ventured much into cloud storage because data is still stored locally. Researchers, academic staff, and students typically use external hard drives, flash disks, CDs, DVDs, emails and Google Drive to store their data. Not many use Dropbox, OneDrive, and other Cloud storage media. However, this trend is risky because the library faces several challenges especially power outages that lead to a computer crash, theft of computer hard drives, and other storage media. There is also a danger of data breaches by unauthorized persons since the repository server is not within the confines of the library. Therefore, cloud storage is an important choice for the library to use in order to mitigate the danger of data loss.

Figure 2 depicts how university libraries provide library services via cloud services. Due to the unreliability of non-web based storage media, university libraries have refocused attention to an alternative; cloud service which is web-based. In providing library services to university faculty, students and researchers; research assets in the form of electronic theses/dissertation, articles, research datasets, research reports are stored in the cloud. It is important to note that cloud services provide advantages like large storage space, data back-up among others which non-web based media does not have. However, alternative storage media (cloud computing) appears to accommodate the concerns of university libraries. Putting in place, security of content, defining accessibility levels, adherence to copyright and legal issues, cloud storage policy, among others, safety of research assets on the cloud service is safer.

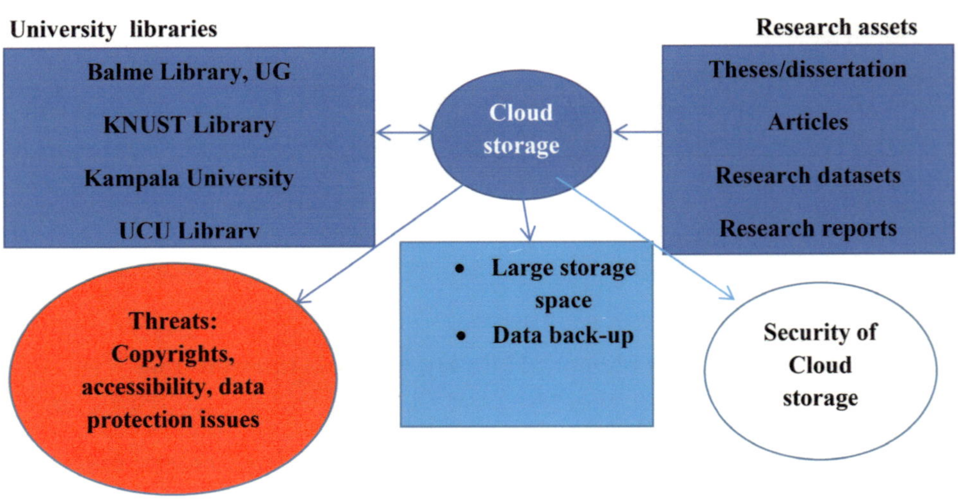

Figure 2.
Cloud computing in university libraries.

6. Summary of key findings

The paper discovered pertinent and important findings which were very vital for drawing a conclusion and informing policy makers.

6.1 Existing storage media for research assets

From the study, it can be concluded that all the sampled academic libraries used magnetic disk drives (hard disk drives) for storing research outputs and assets and optical disks (CD-R and DVD-R).

6.2 Need for digital information storage

From the empirical evidence, it is concluded that information enhances knowledge, which affects behavior, and leads to development warranting its preservation. University libraries have the digital format as text, audio, video, and image which facilitates easy sharing of information. Storage is needed for current and future generation of researchers and academia as a whole. In addition, digital storage makes information easily accessible to users as compared to "analog items". This is due to the ability to easily copy the information on storage devices and carry around. Furthermore, digital storage facilitates the easy sharing of information.

6.3 Risks associated with cloud storage in university libraries

Specifically, copyright law infringement, unauthorized data accessibility, policy issues, the security of content, no interoperable cloud standards were identified as the risks associated with cloud storage in academic libraries.

6.4 Conclusion

Cloud computing offers university libraries improved storage solutions. In the era of IT, the library and information environment face numerous challenges including constant change of storage platforms. Notably, the storage of research output is primary to the functions of university libraries. Thus, there is a need for storage security; as it is a reality in the current technological environment.

In the developed world, some university libraries have already built and managed their own research data centers comparable to the developing world. Indeed, to avoid loss of data integrity, large digital storage in the cloud must be backed up, maintained and re-produced to avoid stress on the local server infrastructure. In conclusion, the opportunities offered by cloud computing via its storage services could ensure that university libraries gain more control over research output.

University libraries must consider investing in cloud infrastructure as it assures large savings or cost effectiveness in operational cost and tech-start-ups [44], paying for what you use and risk transfer and availability [45], scalability, accessibility [4], on-demand service, access to a large network, rapid elasticity and resource sharing [46]. Above all, Gosavi et al. [20] pointed out that libraries are likely to benefit from cloud storage in the area of self-healing, multi-tenancy, linearly scalable, service-oriented, SLA driven, virtualized and flexibility of services.

The paper contributes to knowledge by protecting research data in cloud storage systems. Furthermore, the implication of the findings gives significant input to policymakers, information professionals and future researchers. Finally, with qualitative data, the adopted framework indicates how the security of cloud storage can be implemented successfully.

6.5 Recommendation

The authors recommends the following; security/confidentiality of content, the resilience of librarians, determining access levels and enterprise cloud storage platforms if research output can be secured on the cloud;

6.5.1 Security of content

Content concerns raised by Cave et al. [47] and Genoni [48] require consultation with legislation or the legal office of the academic institution. This is to say that the type of records and length of time for keeping research output must be determined, and policy put in place. In a fast-changing library environment, the technology for storage of research output suffers from obsolescence hence the need for regular back-ups to avoid data loss. Whichever way one considers the issue, storage and access concerns are central, leading to the consideration to make the cloud a viable option.

6.5.2 Resilience of librarians

There is a need for university librarians to maintain the character of resilience and also be adaptive to inevitable and unpredictable changes that occur at an accelerated pace. It is therefore required of librarians to provide a wide variety of information from an equally varied selection of sources and formats through teams (working together) and particularly with the prevalence of cloud use. Since cloud computing enables almost a new streamlined workflow, cooperation through team building or network can be very laudable.

6.5.3 Determining access levels

To overcome the enumerated challenge of unauthorized access to data centers, academic libraries must be concerned with the levels of accessibility; ranging from completely open access to highly private. In securing the content of the research assets on the cloud, different levels of accessibility or privileges must be assigned to the different users within the network. For instance, students, researchers, librarians, users outside the university community must be assigned roles as such.

6.5.4 Enterprise cloud storage platforms

The authors highly recommends the enterprise cloud storage platforms such as Amazon Simple Storage Services (Amazon S3), Google cloud storage and Microsoft Azure. This is because they provide secure, durable, highly-scalable object storage, allows retrieval of any amount of data at any time and high-reliability performance and wide services including operating systems, frameworks, tools, and databases.

Author details

Samuel Owusu-Ansah[1*], Stephen Budu[2] and Rosina Akuokor Anati Budu[3]

1 Balme Library, University of Ghana, Accra, Ghana

2 Presbyterian University College, Abetifi-Kwahu, Ghana

3 Akrofi Christaller Institute of Theology, Mission and Culture, Ghana

*Address all correspondence to: sowusu-ansah@ug.edu.gh;
samsfire55@gmail.com

References

[1] International Data Corporation (IDC). Worldwide cloud IT infrastructure spend grew 21.9% to $29.0 Billion in 2015. 2016. Press release 8 April 2016. http://www.idc.com/getdoc.jsp?containerId=prUS41176716 [Accessed: 17 May 2018]

[2] Alemna AA. An overview of library and information research scene in West Africa. African Journal of Library, Archive and Information Science. 1998;**8**(1):1-12

[3] Shaw KA. Cloud computing for libraries: An economic strategy. In: International Conference on Academic Libraries. 2013. pp. 162-167

[4] Yuvaraj M. Inherent conceptions of cloud computing among library and information science professionals. Library Philosophy and Practice (e-journal). 2015;**1321**

[5] Reese SD. Journalism and Globalization. Sociology Compass. 2010;**4**:344-353. DOI: 10.1111/j.1751-9020.2010.00282.x

[6] Gabridge T. The last mile: The liaison role in curating science and engineering research data. Research Library Issues. A Bimonthly Report from ARL, CNI, and SPARC. 2009;**265**:15-21

[7] Gold A. Cyberinfrastructure, data, and libraries. Part 2: Libraries and the data challenge: Roles and actions for libraries. D-Lib Magazine. 2007;**13**(9/10)

[8] Jones E. Reinventing science librarianship: Themes from the ARL-CNI forum. Research Library Issues: A bimonthly report from ARL, CNI, and SPARC. 2009;**262**:12-17

[9] Witten IH, Bainbridge D. How to Build a Digital Library. San Francisco: Morgan Kaufmann; 2002

[10] Rosenberg AA, Swasey JH, Bowman M. Data mining: Practical machine learning tools and techniques. Burlington: Morgan Kaufmann Publisher; 2006

[11] Lamptey R, Agyen-Gyasi K. The Vision of the Future Academic Libraries in Ghana. Ghana: University Library, Kwame Nkrumah University of Science and Technology; 2015. pp. 1-21. Retrieved from: http://ir.knust.edu.gh/bitstream/123456789/567/1/VISION%20CULD%20Final.pdf [Accessed on: 07/09/2015]

[12] Mayank Y. Cloud libraries. Library Hi Tech News. 2015;**32**(8):19-23

[13] McLeod J, Gormly B. Using the cloud for records storage: Issues of trust. Archival Science. 2017;**17**:349. DOI: 10.1007/s10502-017-9280-5

[14] Karim AJ. The significance of management information systems for enhancing strategic and tactical planning. Journal of Information Systems and Technology Management. 2011;**8**(2)

[15] Taylor J. Organizational culture and the paradox of performance management. Public Performance & Management Review. 2014;**38**(1):7-22

[16] Enakrire R, Baro E. Patterns of information storage and retrieval in university libraries in Nigeria. Library Hi Tech News. 2008;**25**

[17] Conway MA. Autobiographical knowledge and autobiographical memories. In: Rubin DC, editor. Remembering our past: Studies in autobiographical memory. New York, NY, US: Cambridge University Press; 1996. pp. 67-93. DOI: http://dx.doi.org/10.1017/CBO9780511527913.003

[18] Mavodza J. The impact of cloud computing on the future of academic library practices and services. New Library World. 2012;**114**(3/4):132-141

[19] Scale MSE. Cloud computing and collaboration. Library Hi Tech News. 2009;**26**(9):10-13

[20] Gosavi N, Shinde S, Dhakulkar. Use of cloud computing in library and information science field. International Journal of Digital Library Services. 2012;**2**(3):51-106

[21] Li X. The practicality of cloud computing. Library Faculty Publications Paper 11. 2012. Available from: http://digitalcmmons.cacredheart.edu/lirary_staff/11

[22] Han Y. On the clouds: A new way of computing. Information Technology and Libraries. 2010;**29**(2):87-92

[23] Han J, Park K, Pennacchi G. Corporate Taxes and Securitization. The Journal of Finance. 2015;**70**:1287-1321. DOI: 10.1111/jofi.12157

[24] Haris S. Bright Future with Library Clouds, Research Information. 2014. Available from: http://www.researchinformation.info

[25] Kaushik A, Kumar A. Application of cloud computing in libraries. International Journal of Information Dissemination and Technology. 2013;**3**(4):270-273

[26] Teregowda PB, Urgaonkar B, Giles C. Cloud Computing: A digital libraries perspective. IEEE 3rd International Conference on Cloud Computing. 2010:115-122. DOI: 10.1109/CLOUD.2010.49

[27] Saleem A, Tabusum S, Batcha M. Cloud Computing, and Its Applications in Libraries. 2015. Available at: www.academia.edu/11646199/Cloud_Computing_and_Its_Applications_in_Libraries [Accessed: 3 February 2017]

[28] Zainab AN, Chong CY, Chaw LT. Moving a repository of scholarly content to a cloud. Library Hi Tech. 2013;**31**(2):201-215. DOI: 10.1108/07378831311329013

[29] Pandya M. Cloud computing for libraries: A SWOT analysis. In: 8th Convention PLANNER-2012. Gangtok: Sikkim University; 2012. pp. 387-394

[30] Lili L, Buer VB. Reference service evaluation at an African academic library: The user perspective. Library Review. 2015;**64**(8/9):552-566

[31] Liu W, Cai HH. Embracing the shift to cloud computing: Knowledge and skills for systems librarians. OCLC Systems & Services: International Digital Library Perspectives. 2013;**29**(1):22-29

[32] Luo L. Reference librarians' adoption of cloud computing technologies: An exploratory study. Internet Reference Services Quarterly. 2013;**17**(3/4):147-166

[33] McKendrick J. US lags much of world in cloud computing adoption: Study. Forbes.com. 2012. p. 29

[34] De Filippi P, Vieira MS. The commodification of information commons. International Journal of the Commons. 2013 Special Issue: The Knowledge Commons: From Historical Open Science to Digitally Integrated Research Networks

[35] Brownsword, L. (2009) Current Perspectives on Interoperability. Software Engineering Institute, Carnegie Mellon University, 2004. Available from: http://www.sei.cmu.edu/library/abstracts/reports/04tr009.cfm

[36] Zainab AN. Open access repositories and journals for visibility: Implications for Malaysian libraries. Malaysian Journal of Library & Information Science. 2010;**15**(3):97-119

[37] Kitchin R, Dodge M. Code, Space: Software and Everyday Life. London: The MIT Press; 2011

[38] Mitchell E. Using cloud services for library IT infrastructure. Code [4] lib JThenal. 2010;**9**:1-6. Available at: http://jThenal.code4lib.org/articles/2510 [Google Scholar]

[39] Nurnberg PJ, Leggett JJ, McFarland SM. Cloud as infrastructure at the Texas digital library. Journal of Digital Information. 2012;**13**(1)

[40] Yahya F. A security framework to protect data in cloud storage [PhD thesis]. University of Southampton; 2017

[41] Agrawal R, Nyamful C. Challenges of big data storage and management. Global Journal of Information Technology. 2016;**6**(1):01-10

[42] Ji C, Li Y, Qiu W, Awada U, Li K. Big data processing in cloud computing environments. In: Paper Presented at the Proceedings of the International Symposium on Parallel Architectures, Algorithms and Networks, I-SPAN. 2012

[43] Reed DA, Gannon DB, Larus JR. Imagining the future: Thoughts on computing. Computer. 2011;**1**:25-30

[44] Tritt D, Kendrick KD. Impact of cloud computing on librarians at small and rural academic libraries. The Southeastern Librarian. 2014;**62**(3):1-33

[45] Malhotra DR, Jain P. How to choose an economic cloud deployment model. International Journal of Computer Trends & Technology. 2013;**4**(8):2607-2614

[46] Mell P, Grance T. The NIST definition of cloud computing recommendations of the National Institute of Standards and Technology. NIST Special Publication. 2011;**145**:7

[47] Cave J, Robinson N, Kobzar S, Schindler HR. Regulating the cloud: More, less or different regulation and competing for agenda. 2012. Available at: http://ssrn.com/abstract=2031695 [Accessed: 18 Nov. 2017]

[48] Genoni P. Content in institutional repositories: A collection management issue. Library Management. 2004;**25**(6):300-306

Chapter 8

e-Governance and Anti-Corruption War in Africa: The Nigeria Experience

John Sunday Ojo

Abstract

The failure of traditional governance model to provide solution to the contemporary challenges in the public sector has elicited the advent of sophisticated technological response that provides a placid environment for digitalizing public administration across the world. The development of ICTs therefore brings transformation in the way and manner of governing the citizens in Africa and the global community. The transition from governance to e-governance has been considered as a veritable instrument in ensuring transparency, accountability and effective service delivery in the public sector. The potency of ICTs as a supporting tool in curbing the pervasiveness of institutional corruption in Africa is acknowledged. Globally, Nigeria is recognized as one of the most corrupt countries in the world, corruption has been a major problem setting-back the advancement of the country. In taming the obnoxious trend, the recent public sector reform championed by the Nigerian government aimed at providing sanctity in the management and utilization of public resources with the application of ICTs geared toward combating corruption has gained momentum. These institutional reforms provide a platform for e-governance in managing public resources for the benefit of the citizenry. Therefore, this chapter provides a critical examination of e-governance model employed in tackling corruption in Nigerian public sector.

Keywords: corruption, anti-corruption, e-governance, transparency, Africa, Nigeria

1. Introduction

Technological advancements have been credited for playing a significant role in the globalization of trade, communication, economics, politics, culture and life styles. Thus, modern communication technologies or information and communications technologies (ICTs) have been ascribed to not only improving efficiency or productivity in the business world but also with improving the standard of living for the global citizens. One of the strategies is the application of e-governance which entails the use of Information and communication technologies (ICTs) in managing governmental business in the region. Information and communication technologies (ICTs) were the stimuli behind the economic, political and social revolution since the invention of the digital computer (mid-1940s), which were accelerated with the advent of the Internet (the early 90s). Currently, the work of public institutions depends increasingly more on the technical and architectural choices that are made at the level of technology. Most of the principles of good governance are expressed in the roles of e-government; in other words, focus on transparency, openness, citizen participation, effectiveness, efficiency, accountability and the likes [1]. The implementation of e-governance to carry out public services has become a global drive in public administration. It is aimed to develop a robust environment for efficiency in managing public affairs. Since 1960s through 1970s, as information technology emerged, it has been envisaged that the use of modern technology would bring revolution to the way and manner of running governmental businesses. As information communication continues to dominate the space of all organizations especially in the 1980s and 1990s, the political office holders' arrogated recognition for the implementation of information communication technologies (ICTs). Toward the beginning of twenty-first century, the application and the use of ICTs became more popular among the citizens and public officials [2].

In the management of public affairs, corruption has been recognized as one of the dominant problems in promoting economic growth and the welfare of the citizens [3]. Corruption has been acknowledged as a fundamental challenge to good governance and development, habitually embedded in the Global South. Although, it is not only restricted to the region, nonetheless, chronic poverty, conflict and discrimination embedded in developing states are accredited to be responsible for promoting corruption [4]. To combat the increasing level of corruption in developing nations, there are sundry measures put in place to checkmate the syndrome in the public sector. One of such approaches is the e-platform which provides a modern approach to promote anti-corruption agenda, resulting in accountability and transparency in the overall management of public affairs [1]. Thus, e-governance should be considered as a digital mechanism for mitigating corrupt practices in the public sector. Accessible public information for all citizenry can bring about transparency, which limiting public official to demand or accept bribe. One of the major approaches in reducing corruption in the public sector is to limit interaction between the citizens and public officials. This can ensure not only providing accessible information regarding government activities but removing the bureaucratic bottleneck inherent in public administration. Even though e-governance is not the first and last method to be employed for anti-corruption campaign in the public sector, it is being practiced very efficiently in the developed countries and in a few developing countries as well. In general, both developed and developing countries have succeeded much in implementing e-governance, although with relatively uneven attainments. Their efforts to apply the tools and strategies of e-governance have been visible since the last couple of decades. There are countries especially in Asia which have effectively utilized e-governance in reducing corrupt practices and promote good governance. These include The Republic of Korea, Singapore, Japan, Israel, Bahrain, Malaysia, and Sri Lanka [5]. There are other countries in developing nations that have embarked on such a giant stride to fight corruption, albeit they have lagged behind in the effective implementation of e-governance. International organizations such as Transparency International (TI), United Nations Development Programme (UNDP) and World Bank (WB) among others have reinvigorated the need to implement modern technology against the traditional approach in managing government activities [4]. Therefore, this chapter aims to provide the nexus between e-governance and anti-corruption strategies in Africa by exploring the application of e-governance in Nigerian anti-corruption campaign.

2. Transition from governance to e-governance

The traditional concept of governance emerged from the Greek language (κυβερνᾶν), which connotes "to steer". The concept transcends a mere existence of government. In all societies, people employ governance in their daily lives to manage human relationship, interaction and activities. In this context, the use of governance can be metaphorically referred to steering of a ship. Actually, steering of a ship requires navigation toward a particular direction, and this necessitates an existence of seamanship who is saddled with the responsibility of navigating the ship toward certain destination. Thus, governance as a process requires steering of people toward development [6]. Moreover, governance is the capacity of the government to steer society [7]. Nonetheless, to avoid absolutism and self-steering capacity of the appointed/elected leaders, the modern perspective of governance requires peoples' participation in decision making process. The conceptualization of governance in a modern expression stipulates citizens' involvement in public decision making. This makes governance a participatory task by both the ruler and the ruled. The participation of every citizen in allocating and managing public resources defines the modern approach of governance across the globe. According to World Bank, governance implies the management of social, political and economic affairs for sustainable development. The World Bank acknowledged three fundamental elements of governance which include the nature of political regime; the strength and ability of the governments to effectively formulate, implement and discharge responsibilities; and the way and manner the regime in power manage the social and economic resources for development [8]. Therefore, governance entails institutional environment in which individual citizen and the political actors interact and participate in the public decision making. It requires the management of public affairs in a transparent, inclusive, participatory, responsive and accountable manners. Allocating resources and judicious use of these resources for development purpose becomes sacrosanct [9]. Hence, it can be argued that governance articulates and accommodates both the formal and informal actors in public decision making. This is to facilitate equal participation in a more inclusive, transparent and accountable management of public resources guided by the rule of law.

In line with the above, the digital revolution which provides transformation in all human societies delineates the concept of governance from primordial model to a modern perspective which indulges the application of information communication technologies (ICTs) in public decision making. e-Governance as a concept emerged at the inception of 21st century in the management of public sector. There is a paradigm shift from traditionalism to modernism in public administration. The emergence of e-governance therefore radicalized the nature of doing governmental businesses around the world. This opportunity paves way for a new path in conducting and managing public affairs, and this introduced e-services in public sector. Thus, e-governance has been acknowledged as capable of increasing the impact of government on the people [2]. e-Governance implies political and civil operation of government using information and communication technologies (ICTs) by which both the government and citizens interact. It is a strategy employed by the government through the use of ICTs with the aimed of encouraging citizens to participate in governance and make the government more transparent and accountable to the general citizenry [10]. e-Governance entails the use of web-based internet application to interact with citizens, employees, other government agencies, associate or business partners. It encourages participation of every citizen in the government business through technological platform [11]. It has been further argued that "the application of information communication technology (ICT) by the government to enhance accountability, create awareness and ensures transparency in the management of governmental business". It further explained that e-governance can be recognized as a political tool employed by government to promote public oriented services [12]. Therefore, it can be acknowledged from the above view that e-governance appears to be a paradigm shift from traditional model of carrying

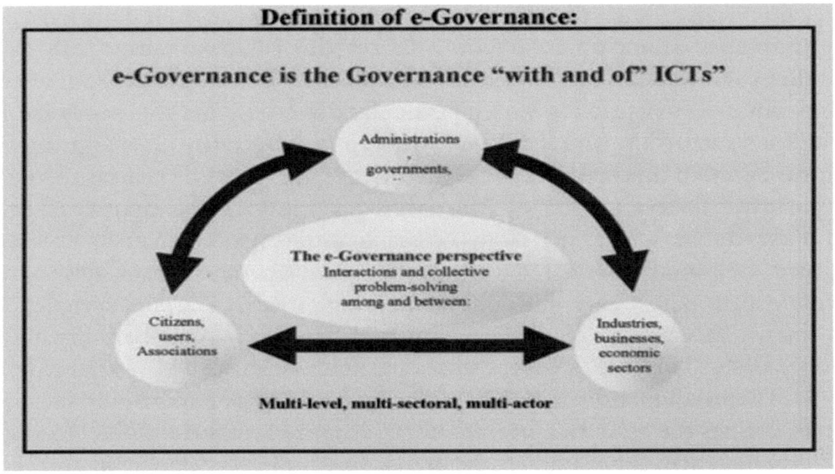

Figure 1.
Definition of e-governance. Source: Misuraca [6].

out governmental businesses which emphasizes compliance with the hierarchical chain of command to the application of internet that allowed the citizens to source for information on government services at their own convenience time which does not necessarily need physical presence in government offices. The primary focus of e-governance is to ensure that citizens have access to government services without direct contact with the public official. The service often delivered through government internet platform. It is believed that the use of internet will reduce the negative impression toward public officials that have been denounced as corrupt. The government websites provide a landscape for the citizens to interact with the government and receive feedback regarding the services provided by the government. In this context, e-governance has the potential to minimize corrupt practices in government establishment [13]. Thus, it can be argued that the major single focus of every e-governance project is to reduce corruption in the management of public affairs across the world.

Many governments around the world have chosen the path of e-governance to manage all government departments and agencies (**Figure 1**). This requires all activities and services provided by the government to be available online such as income tax, customs, property tax and sales tax. The government services are delivered through online platform to the citizenry. e-Governance often necessitates the centralization of data to improve audit and analysis. The inability of the public official to follow online procedures in the conduct of government businesses usually leads to exposure of such act. The citizen complains can also be recounted through e-platform created by government which often requires feedback from the appropriate authority. This environment may facilitate successive strategy to expose corrupt civil servants within the government agencies [11].

3. e-Governance in Africa

In the world over, governments have supported the use of electronic means to provide cost effective services as well as improving the way and manner in which government businesses are carried out. As a result of global acceptance of this strategy, Africa has initiated policies, mechanism and programs toward ensuring effective implementation of e-governance in the region. e-Governance has been one of the key strategies to strengthening democratic governance and as an important instrument in achieving developmental agenda in Africa. The realization and implementation of e-governance has penetrated the social and political space in the region. This makes it possible for all the stakeholders to participate in governance processes. Moreover, the intra-governmental communication provided as a result of the implementation of e-governance makes it easier to reduce the cost of service and also provides wider reach to the general citizens [14]. Among the several strategies for the actualization of e-governance is the creation of websites, linking the citizens with the numerous services provided by the government – Federal, State and Local. The creation of websites therefore becomes interactive space where the citizens and the government can share their views regarding the activities of government while receiving feedbacks on the same platform. Some of the services available to the citizens range from birth certificate, identity card, voter's registration, payment of tenement rate, taxes and levies etc. One of the countries that showcased e-governance capacity in the management of public is South Africa government. The web provides detail information about the government's services. The South African government offers services within the following spectrum: services for the citizens; services for organization; and services for foreign nationals. The services to the citizens revolves around information concerning youth, education and training, retirement and old age, birth, social benefits, relationship, living with disability, parenting, death, citizenship, sports and recreation, law, transport among others. Services implying organizations include import and export, health and safety, labor related issues, business tax, transport, health and safety at work place, intellectual property among others. Information related to foreign national include immigration and working in South Africa. Similar features are also found in Kenya where the issues related to e-civil service, e-business, e-citizenship, e-education, e-taxes and revenue among several other categories [15]. A comparable features are found in Nigeria where activities such as tourism, drug administration, company registration, immigration, tax, investment, driver's license, education, health, housing projects, corruption reporting, national identity management are embedded in government portal. Thus, it can be deduced that e-governance has facilitates citizen's participation as well as reducing cumbersome nature of traditional mode of governance in Africa. The table and figures below reveals the population of internet users and the level of penetration as compared to the rest of the world.

4. Data interpretation and discussion

From the above, **Figure 2** reveals that Asia region constitutes 50.4% of the world internet users, it further shows that Europe has 16.5%. The data also found that Africa has 10.9%, Latin America and Caribbean 10.1%, North America 7.5%, Middle East 3.9% while Oceania and Australia constitutes 0.7%. In this context, Asia has the

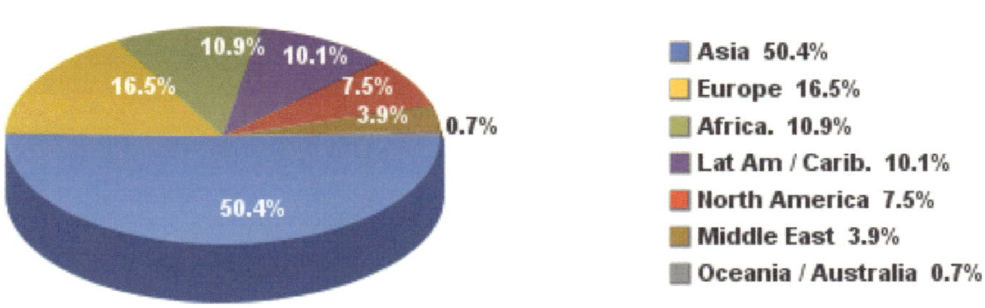

Figure 2.
Internet users in the world by region as at March 2019.

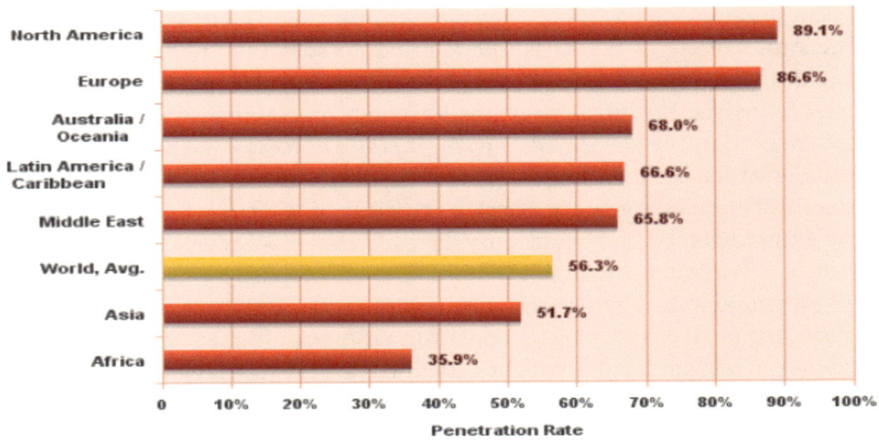

Figure 3.
Internet world penetration rates by geographic regions as at March 2019.

largest number of internet users across the world, followed by Europe. The implication of this as related to e-governance is that the more accessible internet is, the more the people interact with the government on electronic platforms. In **Figure 3**, the data shows internet world penetration rates by geographic regions. The data found that North America 89.1% which constitutes the highest level of penetration globally. This is followed by Europe which has 86.6%. Australia and Oceania has 68.0% while Latin America and Caribbean has 66.8%. The Middle East constitutes 65.8% while the world average is equated to 56.3%. Asia has 51.7% while Africa has 35.9% which is considered as the lowest rate when compared with other regions.

Table 1 shows the Internet Users Statistics for Africa as at March 2019. The table revealed all the countries in Africa with the divergence population, number of

Africa	Population (2019 Est.)	Internet users 31-Dec-2000	Internet users 31-Mar-2019	Penetration (% population)	Internet growth % 2000–2019
Algeria	42,679,018	50,000	21,000,000	49.2	41,900
Angola	31,787,566	30,000	7,078,067	22.3	23.493
Benin	11,801,595	15,000	3,801,758	32.2	25.245
Botswana	2,374,636	15,000	923,528	38.9	6057
Burkina Faso	20,321,560	10,000	3,704,265	18.2	36,942
Burundi	11, 575,964	3000	617,116	5.3	20,470
Cabo Verde	560,349	8000	265,972	47.5	3225
Cameroon	25,312,993	20,000	6,128,422	24.2	30,542
Central African Rep	4,825,711	1500	256,432	5.3	16,995
Chad	15,814,345	1000	768,274	4.9	76,727
Comoros	850,910	1500	130,578	15.3	8605
Congo	5,542,197	500	650,000	11.7	129,900
Congo Dem Rep	86,727,573	500	5,137,271	5.9	1,027,354
Cote d'Ivoire	25,531,083	40,000	6,318,355	25.6	16.246

Africa	Population (2019 Est.)	Internet users 31-Dec-2000	Internet users 31-Mar-2019	Penetration (% population)	Internet growth % 2000–2019
Djibouti	985,690	1400	180,000	18.3	12,757
Egypt	101,168,745	450,000	49,231,493	48.7	10,840
Equatorial Guinea	1,360,104	500	312,704	23.0	62,441
Eritrea	5,309,659	5000	71,000	1.3	1320
Ethiopia	110,135,636	10,000	16,437,811	14.9	164,270
Gabon	2,109,099	15,000	985,492	46.7	6470
Gambia	2,228,075	4000	392,277	17.6	9707
Ghana	30,096,970	30,000	10,110,0001	33.6	33,600
Guinea	13,398,180	8000	1,602,485	12.0	19,931
Guinea-Bissau	1,953,723	1500	120,000	6.1	7900
Kenya	52,214,791	200,000	43,329,434	83.0	21,564
Lesotho	2,292,682	4000	627,860	27.4	15,596
Liberia	4,977,720	500	4,028,418	80.9	805,583
Libya	6,569,864	10,000	3,800,000	57.8	37,900
Madagascar	26,969,642	30,000	1,900,000	7.0	6233
Malawi	19,718,743	15,000	1,828,503	9.3	12,090
Mali	19,689,140	18,800	12,480,176	63.4	66,283
Mauritania	4,661,149	5000	810,000	17.4	16,100
Mauritius	1,271,368	87,000	803,896	63.2	824
Mayotte (FR)	266,380	N/A	107,940	40.5	N/A
Morocco	36,635,156	100,000	22,567,154	61.6	22,467
Mozambique	31,408,823	30,000	5,279,135	16.8	17,497
Namibia	2,641,996	30,000	797,027	30.2	2557
Niger	23,176,691	5000	951,548	4.1	18,931
Nigeria	200,962,417	200,000	111,632,516	55.5	55,716
Reunion (ER)	889,918	130,000	480,000	53.9	269
Rwanda	12,794,412	5000	3,724,678	29.1	74,393
Saint Helena (UK)	4096	N/A	2200	53.7	N/A
Sao Tome & Principe	213,379	6500	57,875	27.1	790
Senegal	16,743,859	40,000	9,749,527	58.2	24,274
Seychelles	95,702	6000	67,119	70.1	1018
Sierra Leone	7,883,123	5000	902,462	11.4	17,949
Somalia	15,636,171	200	1,200,000	7.7	599,900
South Africa	58,065,097	2,400,000	31,185,634	53.7	1199
South Sudan	13,263,184	N/A	2,229,963	16.8	N/A
Sudan	42,514,094	30,000	11,816,570	27.8	39,288
Swaziland	1,415,414	10,000	446,051	31.5	4360

Africa	Population (2019 Est.)	Internet users 31-Dec-2000	Internet users 31-Mar-2019	Penetration (% population)	Internet growth % 2000–2019
Tanzania	60,913,557	115,000	23,000,000	37.8	19,900
Togo	8,186,384	100,000	899,956	11.0	800
Tunisia	11,783,168	100,000	7,898,534	67.0	7798
Uganda	45,711,874	40,000	19,000,000	41.6	47,400
Western Sahara	582,478	N/A	28,000	4.8	N/A
Zambia	18,137,369	20,000	7,248,773	40.0	36,144
Zimbabwe	17,297,495	50,000	6,796,314	39.3	13,492
Total Africa	1320,038,716	4,514,400	474,120,563	35.9	10,402
Rest of World	6,433,444,493	83.0	3,872,441,290	60.2	89.1
World Total	7,753,483,209	100%	4,346,561,853	56.1	100.0

Source: The data was adapted from Internet World Stats. Available at https://www.internetworldstats.com/stats1.htm

Table 1.
Internet users statistics for Africa as at March 2019.

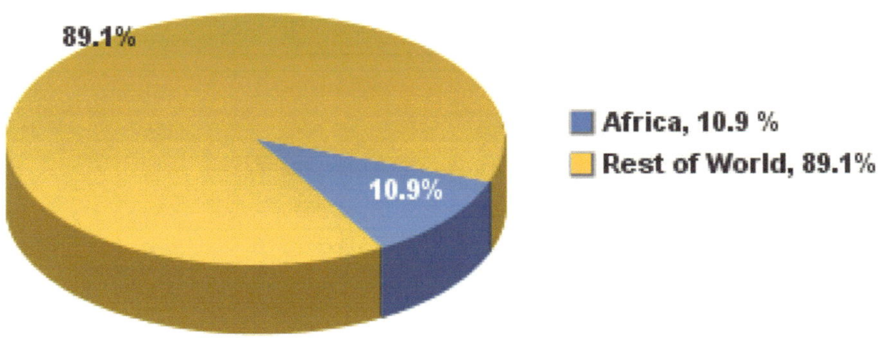

Figure 4.
Internet users in Africa as at March 2019.

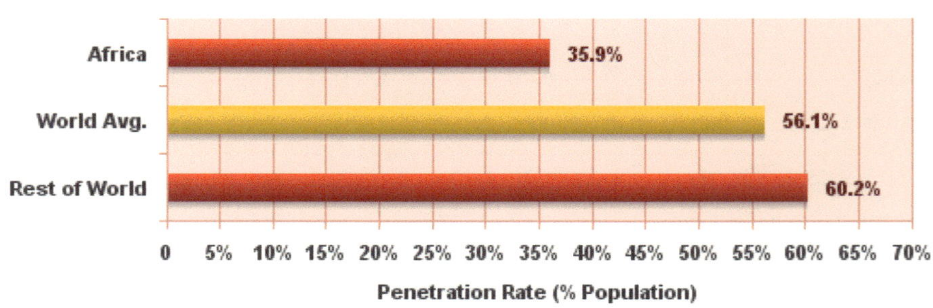

Figure 5.
Penetration rate in Africa as compared to the rest of the world as at March 2019.

internet users, level of penetration and internet growth. In **Figure 4**, the data reveals internet users in Africa as at March 2019. It shows that Africa has 10.9% while the rest of the world constitutes 89.1%. In **Figure 5**, it shows the level of penetration rate in Africa compared to the rest of the world as at March 2019. The data reveals that Africa has 35.9% while world average constitutes 56.1%. The data further shows that

DOI: http://dx.doi.org/10.5772/intechopen.87012

60.2% constitutes the rest of the world. This indicates that, there is a need to reduce the cost of access to digital technology in order to make it affordable for the wider segment of the population. In all, it is acknowledged that Africa has lagged behind in the implementation of digital technology, as indicates in the above data, many African nations have not fully implemented e-governance in managing public affairs. Even though some of these African nations have continually promote the use of ICTs in major government activities, it is believed that the affordability of digital technology will encourage the citizens to employ these platforms to interact more with the government especially regarding governance and anti-corruption campaign.

5. Corruption in Nigerian public sector

Corruption is the abuse of public office for personal gain. Corruption may occur at any levels of government – local, state and national. It can also manifest at the legislative, executive and judicial arms of governments. By and large, corruption exists in every sphere of life, be it public or private establishments. Irrespective of where it occurs, it tends to have a mammoth impact in the general lives of the citizenry especially on the provision of the basic services [16]. Before the introduction of e-governance system, a vast number Nigerian public employees at the local, state and federal levels acknowledged to be terrifically corrupt. The employees in the public sector are recognized to be smart in inflating government contracts and procurements; the system was moribund with ghost worker syndrome in which the names of non-existing public officials were used to pay fraudulent salaries. The committers of such offense initiated a system that bypass banking measure in the payment of salary. In such an environment, they pay public personnel by cash through their ministries cashiers. A huge amount of funds was diverted to the pocket of perpetrators through this channel. Several regimes have contended with this challenge through the inauguration of special staff audit committees, the committee introduced "Table Payment" which requires physical presence of staff. However, this measure was unable to rule out the corrupt elements in the public

Figure 6.
How e-Governance can prevent corruption. Source: Iqbal and Seo [10].

sector [13]. Consequently, the diversion of government revenue and foreign aid into the individual personal purse within the public sector has contributed to the deprivation of some basic social services. It has been estimated that approximately 400 billion was stolen from Nigerian public treasury from 1960 to 1999, between 2005 and 2004, the sum of $182 billion was diverted from the public accounts via illegitimate financial flows. The stolen common wealth meant for educational development, health sector and general infrastructural development was diverted by public office holders for personal aggrandizement [17]. The culture of corruption entrenched in Nigerian public sector has a great impact in the life of the citizens. Its effect on infrastructural development has been acknowledged and its corollary has being experienced by the citizens in their daily engagement (**Figure 6**).

In **Figures 7** and **8**, these platforms showcase the electronic model designed to mitigate corruption in Nigeria. On these platforms, several activities are embedded,

Figure 7.
Independent corrupt practices and other related offenses commission.

Figure 8.
Economic and financial crimes commission.

DOI: http://dx.doi.org/10.5772/intechopen.87012

ranging from corruption reporting, whistleblower channel, number of people prosecuted for corruption, complaint and observation of the citizens against the mandate of anti-corruption institutions, petition among other related activities. These platforms have encouraged the citizens to report several cases of corruption which the anti-corruption institutions have investigated in recent time. Without physical presence of the citizens, corrupt practices can be reported on the website of the Independent Corrupt Practices and Other Related Offenses Commission (ICPC) and Economic and Financial Crimes Commission (EFCC). This effort has yielded an expected outcome following the adoption of e-governance in combatting corruption in Nigerian domain. Therefore, anti-corruption war is considered as government and citizens' struggle, it is believed that if government co-opted the citizens in fighting corruption, corruption will be mitigated to a barest minimum level in the country.

6. e-Governance as a potent weapon for anti-corruption war in Nigeria

The nexus between e-governance and anti-corruption war has been globally pronounced by a number of scholars [12, 18–21]. Information communication technology is considered as a fundamental instrument that can be employed to tame the ugly trend of corruption. ICT is capable of mitigating corruption by promoting good governance, monitoring the activities of government and the governed. The use of electronic measure in the daily governmental business has a great impact in the fight against corruption among other measures put in place by the government in many developing nations of the world [1]. Several efforts have been put forward to prevent corruption by different regimes. The creation of Independent Corrupt Practices Commission (ICPC), Economic and Financial Crimes Commission (EFCC) and Code of Conduct Bureau among others. Despite the creation of different institutions to checkmate the spreading wave of corruption in both the public and private sector in the country, it appears that these measures were not absolutely capable of reducing corrupt practices in Nigerian society. The ineptitude of the above-mentioned measures triggered the development of e-governance scheme which was strengthened to foster transparency in the conduct of public affairs. The implementation of e-governance approach was made possible with the adoption of Information Communication technologies (ICTs) revolution. The implementation process was initiated as a pilot scheme initiated in some ministries and federal parastatals. Following this event, other levels of governments began to adopt this policy to tackle the high level of corruption in the country. The adoption of this policy aimed at block the manifold sources by which public officials employed in siphoning public funds such as double payment of contracts and ghost workers. A huge amount of public funds has been saved by the three tiers of government due to implementation of e-governance project. The philosophy behind this policy is to reduce corruption and increase the level of transparency and accountability in the public and private sectors [22]. It is a by-product of ICT, which is an efficient and effective strategy to tackle corruption idiosyncrasy. Even though e-governance is not the first and last method to curb corruption in the public sector, it is being practiced very efficiently in the developed countries and in a few developing countries as well. In general, developing countries have succeeded much in implementing e-governance. Their efforts to apply the tools and strategies have been discernible since the last couple of decades [4]. The services delivered through electronic means such as tax payment, rate, license are capable of mitigating human error and corruption in the public arena. In an environment where public administration is digitalized, physical contact which has the potential of promoting corrupt practices is mitigated. e-Governance serves the purpose of bringing transparency, accountability and openness to the public service [3]. Nigeria has embarked on numerous strategies in reducing corruption especially through e-governance within the public sector. The public sector reform includes Treasury Single Account (TSA), Biometric Time and Attendance, Integrated Payroll and Personnel Information System (IPPIS), Prepaid Meter. Other initiatives include e-passport, online registration of Joint Admission Matriculation Board (JAMB) by candidates, introduction of computer based examination to reduce exam malpractice, the use of card reader during election, e-reporting of human rights abuse, monthly publishing of local and state governments allocation by the ministry of finance which allows citizens to be aware of how the government use their public resources [22]. According to Davies and Fumega, there are eight kinds of ICT mechanism that can be used to prevent corruption, these include the following:

i. Online services: Platforms offer public self-services that citizens can explore

ii. Transparency portals: These platforms provide periodic government publication of important documents which are accessible to the citizens online;

iii. Open data portals: These platforms offer free access to data sets in machine-readable formats.

iv. Crowd sourced reporting: This entails the citizens to report grievances regarding the activities of government;

v. Online corruption reporting: The platforms that provide opportunity for citizens to report cases;

vi. Online right-to-information requests: platforms that allow citizens to file right-to-information requests.

vii. Issue reporting: platforms that allow citizens to report problems with public services

viii. Service automation: platforms that replace discretionary decision-making by public officials with auditable software processes [23].

In Nigeria, there are policies implemented by the government to curb corruption in the public domain, some of these policies are discussed below:

7. Treasury single accounts (TSA)

The TSA policy was initiated as part of Economic Reform and Governance program in 2004 by the Federal government in Nigeria. TSA is an integrated national bank account in which all the federal government ministries, departments and agencies (MDAs) remit daily government revenue. The implementation of this policy by the national government aimed to block all leakages in Nigerian revenue generating agencies as well as to guarantee judicious use of government resources for the benefit of the citizens. The TSA is one of the components of public financial management (PFM) reforms which was under the third pillar of National Strategy for public service reforms geared toward achieving vision 20:20:20 aimed to address the challenge of ineffective and inefficient cash management in the country. The implementation of this policy necessitates adoption of e-payment system for every single financial transaction, and it came into full force in January 2009. The adoption of this policy aimed at ensuring efficiency, effectiveness, transparency, openness and accountability in the management of the country's financial resources. Although, before the implementation of TSA, the country was moribund with challenge of monitoring all government accounts by the Office of the Accountant General of the Federation (AOGF). The TSA was guided by three fundamental principles: the unification and synchronization of government treasury accounts; only the chief financial agents of the government are saddled with the oversight responsibility in managing cash resources of the government; and there should be comprehensive coverage of both the budgetary and extra-budgetary as well as consolidation government's cash resources [24].

8. Cashless policy

In 2012, the Central Bank of Nigeria (CBN) inaugurated the implementation of cashless policy, projected to eliminate the physical cash flow in the national economy so as to promote e-transaction. The policy was first implemented in Lagos in January 2012. The policy is aimed to reduce the primary cost of banking services; promote modernization and development of Nigeria's payment system in accordance with vision 20:20 goal to be among the top 20 economies by the year 20:20; reducing high cash usage outside the formal sector and therefore provide opportunity for effective management of inflation to promote economic growth; reducing the risk of high handling of cash which encourages robberies and related crimes; and to mitigate systemic leakage that promote corruption. The fundamental objective of this policy is to encourage cashless economy. The cashless economy inculcates the promotion of payment without the involvement of physical cash. It does not totally exclude the use of physical cash from the economy but an economy where the use of cash is reduced to a bearable minimum level. In such a cashless environment, electronic based payments dominate the economy with the use of credit cards or mobile or bank transfer. Some of the components of cashless policy may incorporate e-exchange, e-money, e-brokering, e-finance among others. The primary measures for the cashless policy in Nigeria imply:

i. Daily cumulative cash limit: N500, 000 and N3 million on free cash withdrawals and lodgments by individual and corporate customers respectively. These are upward reviews from the daily cumulative limits of N150, 000 and N1 million set in January, 2012.

ii. Processing fees for withdrawals above limit: 3% for individual and 5% for corporate. These are downward reviews from the respective 10% and 20% fees set in January, 2012.

iii. Processing fees for lodgments above limit: 2% for individual and 3% for corporate. These are downward reviews from the respective 10% and 20% fees set in January, 2012.

iv. Exemptions from processing fees: this applies to accounts operated by ministries, departments and agencies (MDAs) of the federal and state governments, solely for the purpose of revenue collections. Exemptions also extended to embassies, diplomatic missions and multi-lateral and aid-donor agencies, as well as micro finance banks.

The major goal of this policy is to ensure that the larger percentage of Nigerian populace utilizes electronic platform in their daily transaction. Several measures put in place to accomplish the cashless policy include the following methods:

i. Automated Teller Machine (ATM). ATM can be used for the payment of bill, deposit cash, funds transfer, recharge airtime for mobile phone

ii. Internet banking: This can be employed to make instant balance enquiry, funds transfer, payment of application fees and utility bills. Some of the banks require the customers to use token in order to guarantee security and safeguard the account against fraudsters.

iii. Point-of-Sale (POS) Terminals: This can be used to make payment of any transaction made by the customers.

iv. Electronic Transfer: This can be employed to transfer funds electronically from the customer's account to other sources or destinations [25].

9. The government integrated financial management information system (GIFMIS)

This is an integral component of the ERGP initiative targeted toward tackling corruption through information technology. It was implemented to provide support to public resource management by employing integrated and automated mechanism to ensure effective and efficient economic system. The implementation of this policy is aimed to enhance transparency, accountability and cost effective public service delivery. The primary objective of the GIFMIS is to key into computerized financial information system for the federal government to increase:

i. Federal government capacity to effectively control and monitor the general expenditure and receipts of MDAs

ii. The capability to understand the general costs of collections of activities

iii. The ability to exhibit transparency and accountability to the public and partners

iv. The ability to access information on government's economic performance and cash flow

v. Medium term planning through a medium term expenditure framework (MTEF).

vi. Internal controls to identify and prevent possible fraudulent actions

vii. The access to information on financial operation

Ultimately, the focus of the GIFMIS is to support and improve the federal government public financial management performance. The financial management obligations constitutes the overall financial management cycle of government which includes budget preparation, budget execution and financial reporting. GIFMIS aimed to be used in all areas of government budget preparation, execution and general management of financial resources. Additionally, GIFMIS enables effective revenue collection through integrated system of revenue collection especially in some agencies like Federal Inland Revenue Service (FIRS), Customs, Nigerian National Petroleum Corporation (NNPC) and providing effective revenue remittance and transfer to the TSA [24].

10. Conclusion

This chapter has examined e-governance as an effective mechanism in tackling corruption in Nigeria. Although, there have been several efforts by the past political regimes in fighting corruption in the public sector, it appears that some of the efforts put in place were able to yield minimal outcomes. The recent initiative regarding application of information communication technologies (ICTs) in the general management of public affairs has been considered as effective mechanism to complement the effort in curbing corrupt practices in the public sector. This chapter demonstrates that e-governance as a modern technological tool is capable of reducing high level of corruption in the public sector in the country. The current initiative to reduce the level of corruption in Nigeria necessitates the implementation of e-governance strategy which revolves around the Treasury Single Account (TSA), internet banking, limitation in cash withdrawal, the Government Integrated Financial Management Information System, cashless policy among others. These numerous strategies have been utilized to minimize corrupt practices in Nigeria. Some of these strategies have provided opportunity to prevent ghost workers from continuous existence in Nigerian public sector. Although, the adoption of these measures has been producing positive outcomes, however, there are numerous challenges affecting the effective functioning of e-governance especially among the citizens. These challenges range from social arena where there is low literacy level, poor basic education, lack of access to internet by rural populace, lack of feedback, low level of technological adaptation, different languages, shortage of skills, poor IT literacy as well as political aspects that revolves around lack of cyber laws, poor reform agenda, low budget allocation among others. To make the use of ICTs more effective in the public sector, the government needs to ensure training and retraining of its staff in handling ICTs for effective service delivery, provide affordable market for internet users, re-orientate the local community of its benefits and provide enabling environment for effective implementation of e-governance at the local, state and national levels.

Author details

John Sunday Ojo
Erasmus University Rotterdam, Rotterdam, Netherlands

Address all correspondence to: ojs4scholar@gmail.com

References

[1] Lupu D, Lazăr CG. Influence of e-government on the level of corruption in some EU and non-EU states. Procedia Economics and Finance. 2015;**20**:365-371

[2] Fatile J. Electronic governance: Myth or opportunity for Nigerian public administration? International Journal of Academic Research in Business and Social Sciences. 2012;**2**(9):122-140

[3] Mistry JJ, Jalal A. An empirical analysis of the relationship between e-government and corruption. The International Journal of Digital Accounting Research. 2012;**12**:145-176

[4] Kanchana DG, Samarakoon AK. The role of e-governance in curbing public-sector corruptions (a theoretical overview). OUSL Journal. 2018;**13**(1):5-27

[5] UN e-Government Survey. United Nations Department of Economic and Social Affairs; 2014

[6] Misuraca G. e-Governance in Africa, From theory to Action: A Handbook on ICTs for Local Governance. Ottawa, ON Canada: Jointly Published by Africa World Press, Trenton, New Jersey and International Development Research Centre; 2007

[7] Kjar AM. Governance. UK: Polity Press; 2004

[8] World Bank. Governance and Development. Washington DC: World Bank; 1992

[9] UNESCO. Concept of Governance. 2019. Available from: http://www.ibe.unesco.org/en/geqaf/technical-notes/concept-governance

[10] Iqbal MS, Seo J-W. e-Governance as an anti-corruption tool: Korean cases. Journal of Korean Association for Regional Information Society. 2008;**11**(2):51-78

[11] Kaur S. e-Governance-combating corruption strategy. International Journal of Computer Science and Technology. 2015;**6**(1):70-73

[12] Ojo JS. e-Governance: An imperative for sustainable grass root development in Nigeria. Journal of Public Administration and Policy Research. 2014;**6**(4):77-89

[13] Jimoh RG, Longe OB, Ndunagu JN. Information about electronic governance: A tool to curb corruption in Nigeria. Journal The Messenger. 2018;**10**(2):187-194

[14] UNDP. Fight Corruption with e-Government Application. Thailand: United Nation Development Programme; 2009

[15] Onyancha O. e-Governance and e-government in Africa: Webometrician's perspective of the challenges, trends and issues. In: Paper Presented at the Moi University 3rd Annual International Conference; 31 July to 4 August 2007; Kenya

[16] Ear-Dupuy H, Serrat O. Fighting Corruption with ICT: Strengthening Civil Society's Role. Singapore: Asian Development Bank; 2017

[17] Hoffman L, Patel R. Collective Action on Corruption in Nigeria: A Social Norms Approach to Connecting Society and Institutions, Chatam House Report. Royal Institute of International Affairs; 2017

[18] Anderson. e-Government as an anti-corruption strategy. Information Economics and Policy. 2009;**21**:201-210

[19] Shim DC, Eom TH. e-Government and anti-corruption: Empirical analysis

of international data. International Journal of Public Administration. 2008;**31**(3):298-316

[20] Mauro P. Corruption and growth. Quarterly Journal of Economics. 1995;**110**(3):681-712

[21] Garcia-Murillo M, Ortega R. Do e-government initiatives reduce corruption? Available at: SSRN 2012470; 2010

[22] Danfulani J. e-Governance: A Weapon for the Fight Against Corruption in Nigeria. Sahara Reporter, 2013

[23] Davies, Fumega. Mixed Incentives: Adopting ICT Innovations for Transparency, Accountability, and Anti-Corruption. Norway: Chr. Michelsen Institute (U4 Issue 2014:4). p. 38

[24] Odia JO, Odia AA. e-Government and corruption in Nigeria: The case of treasury single account (TSA). Afro Asian Journal of Social Sciences. 2016;**VII**(IV Quarter IV):1-25

[25] Ayoola T. The effect of cashless policy of government on corruption in Nigeria. International Review of Management and Business Research. 2013;**2**(3):682-690

Chapter 9

Crowdsourcing Strategy of Information Society

Tetsuro Saisho

Abstract

In the modern information society, it has become possible to develop new business models that utilize ICT (information and communication technology) in various industrial fields, industries, and business sizes and types. Crowdsourcing is attracting attention as a new business model in this society. In crowdsourcing, business persons are using Internet websites (crowdsourcing platforms) to generate orders for new business activities, such as "order work to an individual" and "require job orders". In addition, through the utilization of crowdsourcing, the lifestyle of workers is also changing with the provision of new job opportunities. In other words, crowdsourcing offers a new working style to a wide range of people. Thus, the spread of crowdsourcing has created new options for how individuals work and live. In the modern information society, crowdsourcing is a new business model that links a business person (orderer) and a worker (contractor) through a website. In this paper, we outline this new business model for the information society, and discuss the current situation of crowdsourcing, which looks to further influence lifestyle changes in the future. We propose a new business model using ICT, and consider the current situation and potentialities of crowdsourcing, which has elements that may result in major changes in the nature of employment in the future.

Keywords: information society, crowdsourcing, matching business, strategic positioning

1. Introduction

In the modern information society, it has become possible to develop new business models that utilize ICT (information and communication technology) in various industrial fields, industries, and business sizes and types.

Business development using ICT involves more than the typical industries from the past. In the creation of new businesses, it can be applied to a wide range of fields including SMEs (small and medium enterprises) and VEs (venture enterprises) [1].

For example, in the manufacturing industry, an inventory management system is introduced to manage product inventory and determine optimal inventory numbers; while, in the distribution industry, a sales management system is introduced to receive orders and deliver products.

With the introduction of ICT into other fields, there is support for design and drafting by CAD (computer aided design). In addition, the introduction of CIM (computer integrated manufacturing) at production sites can manage the entire range of production activities. With the introduction of CALS (commerce at light speed) in the product life cycle, information, from product design and manufacture to settlement, is shared between the customer side and sales side. Further, companies in the finance, education, and service industries, for example, are developing new businesses using cutting-edge ICT.

With the advent of the information society, crowdsourcing is attracting attention as one of the new business models that utilize ICT. Crowdsourcing uses a crowdsourcing platform, typically a web site on the Internet, to allow business persons to place orders for an unspecified number of individuals and recruit contractors. In this way, crowdsourcing is creating new forms of work ordering and employment.

In sum, in the modern information society, crowdsourcing is a new business model that links a business person (orderer) and a worker (contractor) through a website. In this paper, we will outline this new business model for the information society, and discuss the current state of crowdsourcing, which has elements that promise major changes in the nature of employment and lifestyle in the future.

2. Information society and crowdsourcing

2.1 What is crowdsourcing?

Crowdsourcing is a coined word meaning "outsourcing works to an unspecified person (Crowd)" [2]. As such, it is "a mechanism to access an unspecified number of individuals or companies via the Internet to procure the necessary human resources" (**Figure 1**) [3].

In other words, through crowdsourcing, business persons use ICT, and through crowdsourcing platforms (websites) of crowdsourcing site business person (orderer), to a large number of general unspecified people (worker, contractor) it is the business consignment (outsourcing) of procuring workers from the outside by conducting business orders of the company and recruiting of contractors in the business.

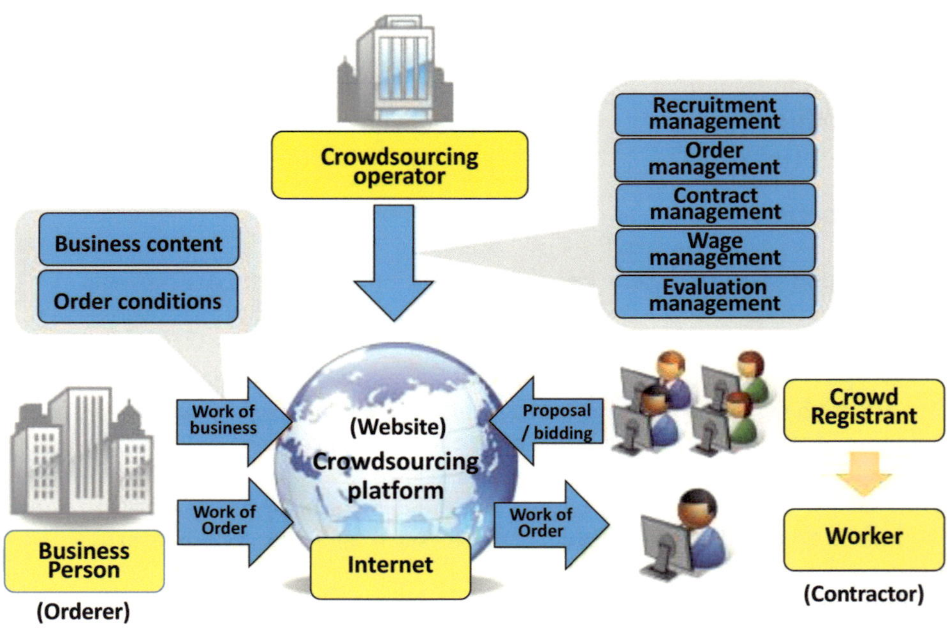

Figure 1.
Crowdsourcing participants and structures. Source: created by author from Ministry of Internal Affairs and Communications (MIAC) (2014), impact on society brought about by rapid evolution of ICT, "Information communication white paper 2014 edition", MIAC.

The respective crowdsourcing relationships between the business person (orderer) and worker (contractor), the crowdsourcing platform of the website (broker), and the flow of business orders, are as follows.

1. The business person uses crowdsourcing in operations where it is difficult for the company alone to respond to a given need, or in operations where external ordering is most efficient.

2. The workers cover a wide range, from individuals without special skills to specialists with well-defined expertise.

3. The business person inputs the work content and order acceptance conditions (work period, work time limit, remuneration, salary, fee, etc.) to the crowdsourcing platform of the website.

4. The worker confirms the work content in the crowdsourcing platform and applies for the specific work.

5. For each task, matching between the business operator and worker is performed, the worker is determined, and a business contract is concluded.

6. The worker is paid a fee after delivering deliverables to the business person.

In the crowdsourcing site business person, the crowdsourcing platform (website) carries out management operations such as defining work content and work location, recruitment management (such as compensation), ordering management, contract management, payment management, and performance and evaluation management.

Crowdsourcing contractors (workers) do not have to work at the business person's facility or at a location near the business person. In other words, workers can work anywhere with Internet access: urban areas, rural areas, sparsely populated areas, islands, overseas, etc. Workers can perform work anytime, anywhere, regardless of the business person's location.

In a 2014 survey involving Freelancers Union, a group for freelancers (workers) in the United States, and Elance-oDesk, a crowdsourcing company, the following elements were identified: "(1) independent contractor: work for a specific project or contract; those who hold a contract", "(2) moonlighter: those who hold a side job from late at night to early morning while holding a regular job", "(3) diversified worker: those who have multiple income sources", "(4) temporary worker: specified by non-regular employment; those who work with their employers", "(5) business owner: have five or fewer employees and work themselves as workers and businesses" [4–6].

The percentage of each element type was: (1) independent contractor (40%), (2) moonlighter (27%), (3) diversified worker (18%), (4) temporary worker (10%), (5) business owner (5%). There are many ratios of independent contractor and moonlighter.

2.2 Crowdsourcing and cloud computing

In recent years, one of the information system configurations that has attracted attention in organizations such as companies is cloud computing, which may on the surface appear similar to crowdsourcing, but is in fact fundamentally different'.

Crowdsourcing outsources work to unspecified people through the Internet; whereas, in cloud computing, computer resources such as software and data related to computer management and use are employed as needed, by means of the Internet [7, 8].

Cloud computing can be roughly divided into (1) SaaS (Software as a Service), (2) PaaS (Platform as a Service), and (3) IaaS (Infrastructure as a Service), according to the types of functions and resources to be provided.

1. SaaS is an Internet-based service that enables remote control of application software having a specific function. SaaS services include online storage and office software that can be used with a web browser.

2. PaaS is a service that enables remote control of the software execution environment via the Internet. PaaS service allows a subscriber to remotely operate an environment in which an OS or middleware has been installed on a computer.

3. IaaS enables remote control of the virtualized computer itself via the Internet. In IaaS service, users prepare, install, and operate all software, such as the OS.

Thus, the "cloud" in cloud computing is a symbol representing a cloud on the other side of a network in a system configuration diagram; whereas, the "crowd" in crowdsourcing refers to an unspecified crowd of people meeting on the Internet platform. "Cloud sourcing", meanwhile, is a specific form of outsourcing in which companies, etc. outsource operations to external vendors and workers.

2.3 Business object of crowdsourcing

Through crowdsourcing, one can rapidly secure workers in-house, such as for software development, including "writing of text", "design of logo and illustration", "programming of the Web and application"; as well as pursue recruitment initiatives based on a specific plan. It is possible to procure and process workers from outside the company in a short time and at a low cost; workers, for example, with specialized skills that are difficult to perform, or a group of workers.

In addition, in crowdsourcing, operations are outsourced widely to professionals and non-professionals. Thus, crowdsourcing is more than the outsourcing of highly specialized tasks demanding specialized skills. Crowdsourcing covers a wide range of projects, including the outsourcing of simple tasks such as data entry and information gathering.

Among the typical crowdsourced task categories are: "Explanation", "Graphics", "DTP (Desk Top Publishing)", "Web Design/Coding", "Content Writing", "Editing of Video/Image", "CMS (Content Management System)", "Blog Homepage", "Application Development", "Web Skills", "Middleware", "Smart Phone Site Development", "Software Development Language", "Environment/Framework", "Data Base", "System Development Technique", "Source Code Management", "Incident Management", "Cloud OS (Operating System)", "Social Media", "3D (Three Dimensions) Technology", "Testing/Operation and Support", "Qualification Skills", "Language/Interpretation/Translation", and "Lawyers"; all of which typically involve specialized work or work requiring special skills.

In addition, among the categories requiring no experience in crowdsourcing, or involving comparatively simple work, are: "Writing (Catch Phrase/Copy Writing, Name Recruitment/Naming, Blog Writing, Review, Article/Content Creation, Document Creation, Editing/Proofreading, Sales Copy/Sales Letter, E-book Production, Mail Magazine Agent/DM (Direct Mail) Creation/Other Writing)", "Task/Work (Data Creation/Input, Question/Question/Testing Test, Internal

Job/Light Work, Various Agency/Call, Photo/Video, Data Classification/Categorization, Other Tasks/Work)", "Office Work/Business Support/Survey (Site Operation/Support, Business Support, Interview, Mystery Shopper, Masking Investigation, Inquiry Response, Email Correspondence, Telephone Support, Sorting of Documents, Business Card Arrangement, Bookkeeping Charges, Secretary Behalf, Easy Investigation, such as the work of other support)"; all of which are typically involved in various fields and industries.

2.4 Forms of employment in crowdsourcing

Thus far, in Japan, the main form of employment of human resources has involved a business entity, such as a company, which bears the primary burden of risk, and engages full-time employees through the conclusion of employment contracts between a business person (orderer/enterprise) and a worker (contractor/individual).

In the information society, however, a wide variety of temporary staffing agencies, worker dispatch companies, job placement agencies, business contractors, etc., have emerged, due to changes in the social environment surrounding businesses, changes in individual needs, and changes in personal lifestyles.

Traditionally, depending on the details of the employment contract between the employer and the worker, full-time employees, contract employees, temporary employees, part-time workers, and part-timers (for whom, in each case, the burden of risk in the employment contract rests on the company, intermediary, agent, etc.) are the typical worker profiles.

However, with the introduction of crowdsourcing, business activities are widely outsourced (ordering) to a large number of generally unspecified individuals. Crowdsourcing thus does not typically involve the execution of work after closing an employment contract between a business person and a worker; but rather the execution of work based on a business contract between a business person and the worker (the risk burden is on the individual).

One result is that, for the modern business person, how to effectively utilize the knowledge resources of specialists outside the organization (out-house group) has become one of the sources of competitiveness in business. Businesses must not depend on themselves alone in the creation of new products and services, using only their own resources and technologies (in-house group). Instead, they must establish core competencies in their areas of activity, and differentiate themselves from other companies.

2.5 Crowdsourcing and competitive advantage

Businesses use specialists and special skills contractors outside the organization, while also carrying out simple operations such as "work with no experience", "work with low expertise", "fine work that can be divided", etc.; thereby making effective use of external resources. In other words, it has become necessary for business persons to outsource their work "when it is necessary", "without direct work", and "with little effort", from their own organization.

Therefore, as a result of the arrival of today's information society, the practical use of ICT is a premise of successful business activity. Unlike in the past, it is often necessary for businesses to make effective use of resources outside the organization, not only in the case of large companies that utilize BPO (business process outsourcing) for most of their operations, but also for SMEs and VEs. To establish their competitive advantage, they must often use resources outside the organization as labor options, for tasks beyond their core competencies.

One of the success factors of modern business is how to make the best use of cheap external labor, and to establish a competitive advantage, each of which is facilitated by crowdsourcing.

In addition, using crowdsourcing, the business person obtains advance information on the prospective worker's quality of work, which is disclosed in advance, through the crowdsourcing platform, enabling him/her to order work after a full and detailed evaluation, including dialog if necessary.

3. Crowdsourcing classifications and functions

3.1 Classification of crowdsourcing

Crowdsourcing modalities can be classified in various ways, and as yet no clear delineation has been made. Thus, here, for practical purposes, they have been roughly classified, based on elements such as the work content and compensation assigned in placing an order, into three types: (1) platform, (2) reward, and (3) order [9].

Hereafter, we will discuss crowdsourcing in these terms.

Crowdsourcing organization in the US, crowdsorcing.org, from the work of crowdsourcing platform handling, (a) cloud labor: simple work requiring high-level skills, (b) crowd creativity: leverage creative talents around the world through the internet, (c) crowdfunding: raise funds for new projects and company establishment from a large number of unspecified people, (d) distributed knowledge: collect, build, and share knowledge assets and information, (e) open innovation: use ideas from individuals outside the organization, are classified into five [10].

3.1.1 Platform type

Platform type crowdsourcing is based on the specific forms of crowdsourcing platform that are provided for workers, and is also referred to as the business classification type of crowdsourcing. The platforms can be classified into three types: (a) integrated platform, (b) sector specific platform, and (c) research and development platform.

3.1.1.1 Integrated platform type

In information system development, crowdsourcing work may include things such as "basic design", "detailed design", "programming", "unit test", "connection test", and "system test". Other information system fields include "website creation", "Web design", "EC (Electronic Commerce) site and net shop construction", "application and smart phone development", "hardware design and development", and "project and maintenance, operation management".

In manufacturing, etc. crowdsourcing may for example include "design work", such as designs and patterns plans and layout creation. In addition, in business writing, etc. there is "copy writing" which aims at advertising copy that is easy and interesting for the user to read, and "Web writing", involving the generation of website text.

In the integrated platform type, there is a wide range of work, including "photograph and video shooting", "data input", "questionnaire", "interview and question", "answering machine", "various agents", "check and inspection, investigation", "office work and secretarial", "accounting and finance", and "human resources and payroll", etc.

As described above, in the integrated type, a business person and worker collaborate in the execution of various orders by handling a wide range of work types.

3.1.1.2 Sector specific platform type

In the fields of logos, banners, and illustrations, crowdsourced work may include "logo creation", "banner creation", "illustration creation", and "game illustration creation"; while other printed matter and DTP design fields include work such as "leaflet creation", "business card creation", "brochure creation", "poster creation", and "packing and package design".

The character, icon, and animation fields include work such as "recruitment of character design", "manga and animation production", "icon creation", and "goods production and novelty". In addition, the map, signboard, and infographic fields include work such as "map making and guide map creation", "signboard design", and "infographics". The POP (Point Of Purchase), menu, and seal fields include work such as "seal and label design" and "POP design".

In the sector specific type, there are operations that limit the types of jobs to specific and often specialized fields such as "binding, book design" and "design and production of CD (Compact Disc) and DVD (Digital Versatile Disc) jackets". As described above, the sector specific type is a system in which a business person and a worker receive and place orders of limited operations by handling specialized operations.

3.1.1.3 Research and development platform type

Business persons use crowdsourcing to address a wide range of issues around the world, in order to respond to their own management, and other, challenges. As a worker (specialist in research, engineering, etc.), it is possible to address various issues, regardless of the time or place.

In the research and development (R&D) platform type, when an enterprise does not have an R&D department, or wishes to accelerate R&D, the company can broadcast its technical issues worldwide and make a broad search outside its own technological and developmental capabilities. In R and D type work, not only researchers at universities and research institutes, but also general inventors and researchers at companies, make technical proposals for the tasks on offer.

The R&D platform type improves the speed of solving technical issues in the enterprise, and facilitates the technical proposals of researchers and engineers. As described above, the R&D platform type involves a business person and workers receive and submit a technical proposal by handling the work of problem solution and proposal.

3.1.2 Payment type

In payment type crowdsourcing, a worker receives compensation as an economic reward. The payment types can be classified into four groups: (a) gift payment, (b) point payment, (c) payment-agency payment, and (d) cash payment.

3.1.2.1 Gift payment

The gift payment worker receives a reward as an economic compensation, such as an Amazon gift or an iTunes gift. Amazon gifts can be used to purchase products that are handled by Amazon.com, Amazon.co.jp, Javari.jp, etc. ITunes gifts can also be used to purchase songs, albums, playlists, audiobooks, music videos and movies in the iTunes Store and the App Store.

3.1.2.2 Point payment

Point payment workers receive, as economic compensation, rewards from various services at points such as PeX Japan (Japan's largest point exchange website), PointExchange Japan (Currently, RealPay), T-POINT Japan (Strategic Comprehensive Alliance, Yahoo! JAPAN, etc.), etc. The points can be used for electronic money, gifts, miles, goods, and various other point types.

3.1.2.3 Payment-agency payment

A payment-agency payment worker receives, as financial compensation, a reward through one of the world's leading payment-agency companies such as Paypal (Singapore), Alipay (China), Pay-easy (Japan), Linepay (Japan), Google pay (US), R pay (Japan), etc. A credit card is required to use the payment-agency method.

3.1.2.4 Cash payment

Cash payment workers receive compensation directly by bank transfer, with cash as financial compensation.

3.1.3 Order payment type

Order payment crowdsourcing is a distinct work order structure. The order types can be classified into four groups: (a) micro task payment (fixed salary), (b) contest competition payment (fixed salary), (c) project payment (fixed salary), and (d) project payment (hourly salary).

3.1.3.1 Micro task payment type (fixed salary)

Micro task payment crowdsourcing consigns small units of work at a low price, such as data entry by workers, or questionnaires, explanatory text creation, and list creation. In this order type, the work content is subdivided, and a supplier is determined for each sub-unit.

3.1.3.2 Contest competition payment (fixed salary)

Contest competition payment crowdsourcing recruits work in a competitive format, such as design, naming, or catch phrase creation, and chooses from among the best. In this order type, a large number of works, ideas, plans, etc., are collected, and the best are selected from among them.

3.1.3.3 Project payment (fixed salary)

In project payment crowdsourcing the supplier is selected in response to a proposal from a worker, for work such as application development or website creation. In this order type, information on the quality of the worker is obtained in advance, and the work is ordered based on judgments of skills, evaluations, comments, etc. In addition, business estimates are solicited and bid for (price, delivery and content, etc.), on the basis of which the supplier is determined.

3.1.3.4 Project payment (hourly salary)

In this order type, crowdsourcing provides time management software for the respective worker's PC, and pays rewards on a time conversion basis for work done. In this case, the worker is paid regardless of whether there are deliverables or not.

3.2 Crowdsourcing features

The relationship between a business person (orderer) and a worker (contractor) in crowdsourcing is typically restricted to exchange via a website (crowdsourcing platform) on the Internet. Crowdsourcing is not a business model that involves work requests to specific regions or individuals, because it is based on work orders for an unspecified number of people or a recruitment of contractors.

With the crowdsourcing system, workers can view and confirm the work content from the projects (work information) provided to the crowdsourcing platform, and then select the work of their choosing, performing it in their own chosen place and time, and with considerable latitude in terms of how much work they choose to do. Then, the worker obtains an "Amazon gift voucher", "iTunes gift", "PeX", "Point Exchange", "T-POINT", "cash", etc., as remuneration for the work.

The crowdsourcing platform (website) thus intervenes between the business person (orderer) and worker (contractor). The platform lists the information, work content, and working conditions, as defined by the business person, and provides them directly and widely to prospective workers.

In addition, the business evaluation of the workers related to the project is performed by the business operator who has placed an order and comments on the crowdsourcing platform. The worker's work history and work contents are screened.

Therefore, we see the emergence of a new kind of business model, wherein the business person can preliminarily evaluate a prospective worker's work quality, select the worker most suitable for the work content and cost, and efficiently place the work order.

Hitherto, in the relationship between a business person (orderer) and worker (contractor), it has been difficult to make such preliminary evaluations before outsourcing the task. However, with the use of crowdsourcing, information about the work quality of the prospective worker (crowdsourcing experienced) can be obtained in advance via the crowdsourcing platform.

The business person can thus place an order for a task, having determined the skills, evaluations, reviews, and the like, of the prospective worker, based on the prior information. Therefore, it is possible to reduce the risk of mismatching based on insufficient preliminary evaluation along conventional lines.

3.3 Typical features of crowdsourcing

The main functions of the crowdsourcing platform are: (1) contact and reporting, (2) business item matching, (3) contract progress and settlement management, (4) business follow-up, (5) business scouting, and (6) administrator management [11]. In each function, matching between a business person and a worker is performed before placing an order for work or recruiting a contractor.

3.3.1 Contact and reporting

In this function, the platform sends an email from the business person to a registered email address, regarding information such as business content and order acceptance conditions. Also, in the text, information can be distributed quickly and efficiently using a fixed format.

3.3.2 Business item matching

This function allows the worker themself to search for work and apply directly for the desired work content. In addition, when the business person places an order for work, it is possible to quickly search for a worker who wants to receive a work order, and to order the work quickly and efficiently.

3.3.3 Contract progress and settlement management

This function performs progress management and settlement management of the project item after the contract has been concluded between a business person (orderer) and worker (contractor). In addition, the payment of compensation to workers can be managed quickly and efficiently through the intermediation of the crowdsourcing platform.

3.3.4 Business follow-up

This function allows the business person and worker to mutually confirm the work content ordered and the work evaluation. Also, both the business person and worker can quickly and efficiently confirm the necessary information.

3.3.5 Business scouting

This function allows the business person to approach the worker directly by referring to information such as the worker's public profile and task evaluation. In addition, the business person can rapidly and efficiently initiate operations by directly approaching workers.

3.3.6 Administrator management

This function can set a fee for prepayment or additional payment of compensation to the worker. In addition, the function enables stable cash management in terms of the constructed business model, including the management of stable funds on the part of both business person and worker.

Thus, once again, crowdsourcing using ICT enables a business model that achieves optimal and rapid matching between a business person (orderer) and worker (contractor), which has traditionally been difficult to achieve.

4. Current state of crowdsourcing

4.1 Features of crowdsourcing

By using crowdsourcing to carry out work, a business person can (1) assign tasks to workers in a timely and appropriate manner, (2) experts and special skills immediately it is possible to order work from the owned worker, (3) if necessary, secure a large number of workers inexpensively. In addition, crowdsourcing makes it possible to quickly and inexpensively evaluate the labor force outside the organization by judging its expertise and skills; and to easily utilize this labor force.

For example, in the case of ordering work by crowdsourcing, it is possible to match the business person (orderer) and worker (contractor) in a minimum of 15 minutes, in a timely and appropriate manner. Also, in terms of price, it is possible to place an order with a conventional business person at a cost of one tenth to one half. Furthermore, in the case of assigning work to specialists or highly skilled workers, it is possible to use split ordering, while checking the quality of workers online. Therefore, business persons can achieve significant cost reductions in their business operations [9].

Traditionally, business persons "outsource work", "adopt recruitment of mid-career labor", "search for work consignment workers", "conclude employment contract with workers", and "time spent on that", etc.; all of which typically involve significant costs before the start of operations. In addition, after contracting with the worker, the business person must factor in "expenses for time spent in fulfilling the contract (coordination costs)", as well as "expenses to ensure that the contracted worker properly executes the contract content (motivation costs)". However, crowdsourcing enables business persons to execute operations without these cost problems, and thereby enhance their competitive advantage.

At the same time, the worker can scrutinize the work information, such as "work place", "work content", "work amount", "work time", "work experience required", "special skills required", "remuneration", etc.; and can themselves control the amount of work directly. For example, full-time homemakers can pursue domestic work and hobbies of interest during intervals of daily work, such as housework, childcare and nursing care, while also receiving outsourced orders for tasks demanding their particular skills. Thus, various forms of crowdsourcing have been developed in response to the needs of both business persons (orderers) and workers (contractors); and currently, many companies provide crowdsourcing services throughout the world, including Japan.

In crowdsourcing, business persons provide business services such as:

a. "System development, design, writing, photograph and video photography, data entry, questionnaire, interview, answering machine, various agency, check, office work, secretary, etc.", which describe a wide range of operations without limiting the field;

b. "Brochure, catalog, flyer, poster, company information, business card, post-card, logo mark production, etc.";

c. "Announce the company's business issues to the world via the website, and seek outside third parties for planning, technology, and development capabilities not possessed by the company." In addition, there are start-up initiatives such as "a business person that develops a single-focus enterprise" and "a business person that develops a multi-focus enterprise".

4.2 Crowdsourcing platform positioning strategy

Crowdsourcing platforms are typically characterized by the work they handle. Thus, a matrix diagram may be configured based on the "work content", "remuneration type, etc." ordered by the crowdsourcing platform. In the matrix chart, the vertical direction (Y-axis) ranges from "Many Items Business" to "Specific Business", and the horizontal direction (X-axis) from "Low Specialization Business, Simple Business" to "High Specialization Business, High Skill Business". In this way, the diagram can be utilized to locate the full range of crowdsourcing business types [12] (**Figures 2** and **3**).

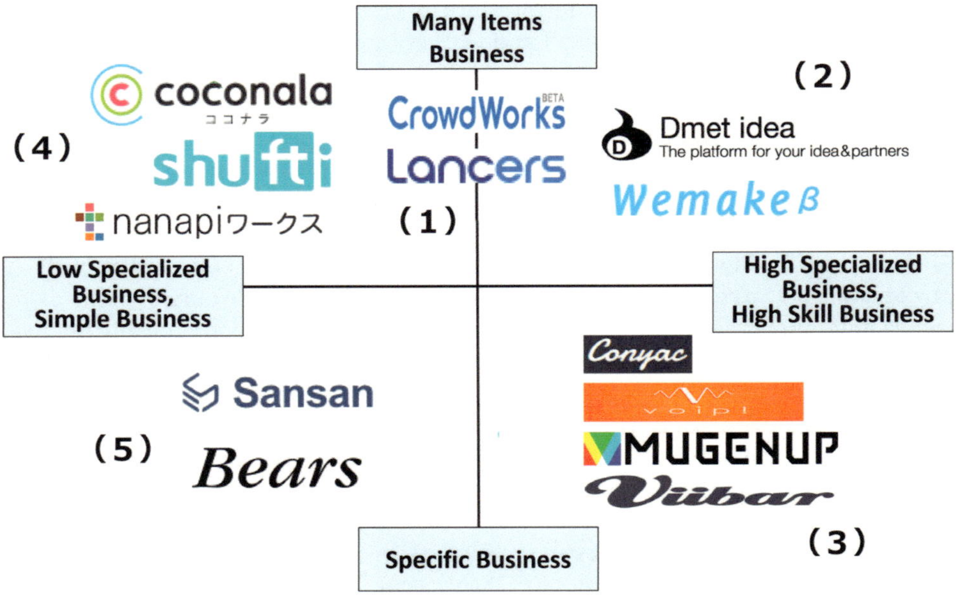

Figure 2.
Positioning strategy for Japanese crowdsourcing companies. Source: created by author.

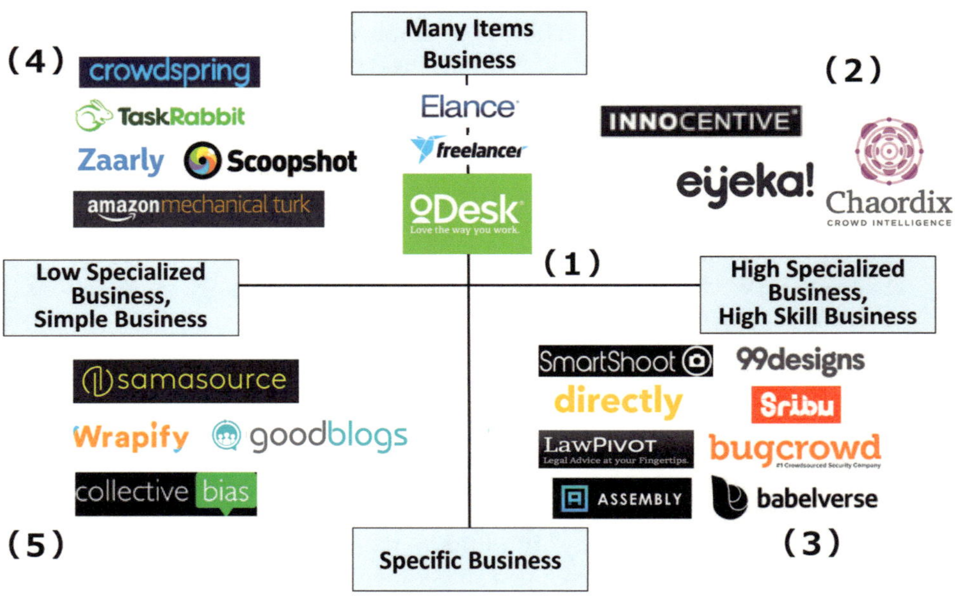

Figure 3.
Positioning strategy for other countries crowdsourcing companies. Source: created by author.

There are five major crowdsourcing platform sectors locatable in the matrix: "(1) Many items business type (general function type)", "(2) Many items business type + high specialization business type (multifunctional type: advanced expertise type)", "(3) Specific business type + high specialization business type (specific function type: advanced functions type)", "(4) Many items business type + low specialization business type (multifunctional type: low expertise type)", "(5) Specific work type + low specialty work type (specific functions type: low functions type)".

4.2.1 Many items business type (general function type)

The crowdsourcing platforms in this service sector are located at the top center of the matrix chart (**Figures 2** and **3**). These platforms differentiate their services from those of other companies by positioning themselves in terms of a "General Function Type". Work types range from "highly skilled work + specialization work" such as programming, writing, translation, web design, etc., to "low skill work + low specialization work" such as simple text creation and data entry provide a wide range of operations.

Major crowdsourcing platforms of this type include "Lancers" and "Cloud Works" in Japan; "Elance" and "oDesk" in the US; "Freelancer" in Australia; and "Witmart" in China.

4.2.2 Many items business type + high specialization business type (multifunctional type: advanced expertise type)

The crowdsourcing platforms in this service sector are located at the top right of the matrix chart (**Figures 2** and **3**). The platforms differentiate their services as "High Specialization + Multifunctional "; and typically offer "many items work" and "work that requires highly specialized skills", for services that combine unique product and service development and existing services.

Major crowdsourcing platforms here include "Dmet idea" (realization of various ideas) and "Wemake" (cardboard furniture using open design) in Japan; "InnoCentive" (expert solutions) in the US; "eYeka" (Ideas competition) in France; and "Chaordix" (question and recruitment of ideas) in Canada.

4.2.3 Specific business type + high specialization business type (specific function type: advanced functions type)

The platforms in this field are located at the bottom right of the matrix chart (**Figures 2** and **3**), and differentiate their services as "High Specialization + Specific Function", offering "highly specialized work" and "work that requires highly specialized skills"; for example, professional photo and video production, real-time interpretation through mobile terminals, order creation of logos and designs in a competitive manner, external blogger requests for company blog writing.

Major crowdsourcing platforms here include "Conyac" (translation), "Voip!" (photograph and video shooting), and "MUGENUP" (illustration production) in Japan; "SmartShoot" (photography, video photography), "99 designs" (design competition), "Directly" (customer care), "Babelverse" (interpretation), "LawPivot" (lawyer), "Bugcrowd" (bug detection), "Assembly" (collaborative development application), and "GoodBlogs" (blog writing) in the US; and "sribu" (design contest) in Indonesia.

4.2.4 Many items business type + low specialization business type (multifunctional type: low expertise type)

The platforms in this service sector are located at the top left of the matrix chart (**Figures 2** and **3**), and differentiate their services as "Low Specialization + Multifunctional", offering "many items work" and "low specialization, low skills work", from simple housework and agency services to office work such as online questionnaires and data entry, and the provision of free-market buying and selling.

Major crowdsourcing platforms here include "coconala" (free market), "nanapi works" (article writing), and "shufti" (office work) in Japan; "TaskRabbit" (housework/use agency), "Zaarly" (housework/use agent), and "Amazon Mechanical Turk" (input work/questionnaire) in the US; and "Scoopshot" (photography service) in Finland.

4.2.5 Specified work type + low specialty work type (specified functions type: low functions type)

The platforms in this service area are located at the lower left of the matrix chart (**Figures 2** and **3**), and differentiate their services as "Low Specialization + Specific Functions", offering "specialized in specific work" and "low specialization, low skills work" such as mental support for women, training and creation of employment for IT skills, and advertisement with markers.

Major crowdsourcing platforms here include "Bears" (support for women's love), "Sansan" (business card management) in Japan; and "Samasource" (IT skills training and employment opportunities for poor women and young people), "Wrapify" (drivers who want to earn advertising revenue with their own cars), "Good Blogs" (blog writing), and "Collective Bias" (influencer marketing) in the US.

4.3 Advantages and disadvantages of crowdsourcing

In the crowdsourcing business model, the following are the main advantages and disadvantages to (1) the business person (orderer) who outsources the work and (2) the worker (contractor) who accepts the work [13–16].

4.3.1 Advantages and disadvantages to the business person.

4.3.1.1 Advantages to the business person

First, business persons can use crowdsourcing to rapidly and efficiently search for optimal workers. Hitherto, hiring outside the organization (employee recruitment) required time-consuming procedures such as recruitment advertisement, resume submission, interviewing, hiring, a trial period, and regular hiring. In addition, it was impossible to properly judge the quality (such as ability and conduct) of the human resources until sometime after employment. However, through crowdsourcing, the business person can obtain information on the evaluations, reputation, etc. of prospective workers in advance, and can quickly outsource the work according to the work content.

Second, business persons can reduce the cost of doing business by using crowdsourcing. Hitherto, employment involving full-time employees incurred costs (personnel expenses such as salary and social insurance expenses) regardless of how much business was actually conducted; and for business persons, the costs for full-time employees were incurred for everything from simple, labor-intensive operations to highly skilled, complex operations.

Third, traditionally, it was necessary to allocate personnel depending on the type of industry, in preparation for major changes in the volume of work, such as upsurges in business activity or activity that typically occurred suddenly and only at the end of the month. However, ideally, business persons should access human resources as needed, and this is enabled by crowdsourcing.

Finally, business persons can use collaborative sourcing to collaborate with other business entities. Hitherto, hiring full-time employees has been in charge of business (projects), performing duties as individual business within an organization or as group business. However, with crowdsourcing, business persons can divide the work into several sub-tasks, and order only part(s) of the total work content from respective workers. And in the case of many work-orders, it is possible to perform effective and efficient operations by leaving some or all of the work in a specific field of expertise to crowdsourced workers.

4.3.1.2 Disadvantages to the business person

First, business persons must ensure information security when using crowdsourcing. In crowdsourcing, a business person consigns work to a worker outside the organization (business order); and in the assignment of work, successful operation may require the disclosure of confidential company information, such as ideas, plans, business schemes, and know-how, as well as proprietary customer information, to the crowdsourced workers. Therefore, confidential information may be leaked through the workers. Thus, when outsourcing work to a worker, it is necessary to decide, before generating an order, how to securely handle the relevant confidential or personal information.

Second, business persons must evaluate and manage work quality, intermediately through the crowdsourcing platform. The business person must, for example, confirm that the task is performed on time and with the expected content. Workers, meanwhile, need not worry about the outcome of the delivered work in continuous outsourcing based on past performance and evaluations. This problem is compounded by the fact that, in practice, a business person will have 10 or more simultaneous work orders in progress, making it difficult to ensure assignment to specific workers. Thus, as it is difficult to assign work based on past results, it is typically necessary to confirm each new worker's skills, past work content, and evaluations.

Finally, the business person incurs a fee for using crowdsourcing, as levied by the Internet website (crowdsourcing platform) that mediates between them and the worker. Crowdsourcing platforms typically have specialized areas such as system development, design, image processing, data entry, etc., with varying fees. Therefore, it is important for business persons to make a careful selection of the optimal crowdsourcing platform for their needs.

4.3.2 Advantages and disadvantages to the worker

4.3.2.1 Advantages to the worker

First, workers can use crowdsourcing to make efficient use of their time. Workers can manage the start, end, and rest time for each task, and effectively develop and apply time management skills. Thus, for example, housewives can take advantage of the gaps in housework and parenting; students, the gaps in their class schedule; and company employees, the gaps in their regular employment schedule. In addition, workers can receive orders for tasks that can be completed in a short amount of time, and works for recruiting ideas.

Second, workers can use crowdsourcing to freely determine their working conditions. Workers can work anywhere, regardless of the place of work or building. Hitherto, in the employment of full-time employees, it was necessary to carry out work in a determinate location, time zone, schedule, building, etc. However, in crowdsourcing, workers can work anytime and anywhere (urban, rural, depopulated areas, islands, overseas, and regions, sites, buildings, places, etc., the location does not matter.) as long as they have an Internet environment.

Third, workers can freely choose desirable work content, such as their favorite type of tasks, work that they specialize in, and work requiring skills they possess. In the past, the hiring of full-time employees had to be carried out in a group based on work orders from superiors to subordinates in the organization. With crowdsourcing, however, workers can be selective in their choice of tasks, choosing those with favorable content or of especial interest, those that make use of sophisticated skills, enjoyable tasks, and tasks well-suited to their abilities.

Finally, workers can use crowdsourcing to avoid troublesome human relationships. In crowdsourcing, email is the primary form of communication between the business person and worker; and since workers do not have to meet the employer directly in order to carry out their work, no extra communication is required. Therefore, troublesome relationships can be avoided, and it is not necessary to have in-person interactions, secure space for them, and pay the transportation cost to get there.

4.3.2.2 Disadvantages to the worker

First, generally, crowdsourced compensation is relatively less than that of full-time employees. One of the advantages of crowdsourcing for businesses is that they can place orders at a lower cost than the cost of conventional full-time employees. Also, though a multifaceted order may be required, the work can typically be subdivided, making it possible to order only a part(s) of the larger project. Therefore, even highly skilled jobs often have relatively low rewards, and workers must compensate by increasing the number of tasks they take on, in order to achieve their financial goals.

Second, workers must confirm the character and integrity of the respective business person through the mediation of the crowdsourcing platform, with communication between the business person and worker basically conducted by email. Workers may thus be hampered in their evaluation, because there is no in-person meeting. In addition, some business persons may, for example, "demand work that is not included in the contract", "become silent while the project is in progress", or "try to avoid paying compensation." Therefore, when actually receiving an order for work, it is necessary for the worker to confirm the order content, rules, etc., and to confirm the integrity of the business person in advance, based on the latter's work-order history.

Third, workers must ensure that their use of the platform is safe and secure. In recent years, advanced cyber-attacks have become a major threat, and attacks from malicious third parties targeting specific organizations are being conducted. If the response to such advanced cyber-attacks is neglected, there is a possibility that "the confidential information of workers and business is leaked", "a fictitious order is placed", and/or "untraceable damage occurs". Therefore, when using a crowdsourcing platform, it is necessary for workers to confirm in advance the relevant BCP (business continuity planning), etc. for any possible failure or trouble in the information system.

4.4 Crowdsourcing market size

In today's information-oriented society, the size of the crowdsourcing market in order for businesses (large enterprises, small and medium enterprises, venture companies, general incorporated associations, NPO corporations, etc.) to establish a competitive advantage in various business fields Are increasing [17–19].

The size of the Japanese crowdsourcing market is 4.4 billion yen in fiscal 2011, 10.66 billion yen in fiscal 2012 (up 242.3% year-on-year), 21.5 billion yen in fiscal 2013 (up 201.7% year-on-year), and 40.8 billion yen expected in fiscal 2014 (189.8% year-on-year) FY2015 forecast of yen 65.0 billion (up 159.3% year-on-year), FY2016 forecast of yen 95.0 billion (up 146.1% year-on-year), FY2017 forecast of yen 135.0 billion (up 142.1% year-on-year), FY2018 forecast of 1820 It has grown rapidly to 100 million yen (up 134.8% from the previous year) [17–19].

A feature of the Japanese crowdsourcing market is that crowdsourcing services are mainly focused on the Japanese market. This is because in the Japanese market, business communication between operators and workers via crowdsourcing platforms is conducted only in Japanese.

Currently, crowdsourcing market in the Japanese culture is limited, and saturation is expected in the near future, and permanent growth of the Japanese market cannot be expected. Therefore, it is necessary to disseminate the Japanese market in the future to unexplored business fields and required workers.

In addition, the Japanese crowdsourcing market can be expected to gain new market growth potential by responding to the global market, including multilingual crowdsourcing platforms.

On the other hand, the scale of the overseas crowdsourcing market has grown to Yen 13.1 billion in 2009, Yen 17.4 billion in 2010 (up 132.8% from the previous year), and Yen 28.9 billion in 2011 (up 166.1% from the previous year) [17–19].

The overseas market size data is the total order amount of work for 15 member companies of Crowdsourcing.org, a US organization related to crowdsourcing, and does not include the order amount of major service companies such as InnoCentive and Amazon Mechanical Turk.

A feature of the overseas crowdsourcing market is that business communication between operators and workers via the crowdsourcing platform is basically conducted in English. In addition, the crowdsourcing platform on the Internet can be used “anytime”, “anywhere”, “anyone”, and “anyone” as long as there is an Internet environment.

Therefore, the crowdsourcing platform is large in size for the global market including the Japanese market. The crowdsourcing platform users are operators and workers around the world. In addition, there are many workers, and it offers high growth potential by providing various services with cheap rewards [20–22].

5. Significance of crowdsourcing strategy

The Crowdsourcing Strategy in the information society has the following three corporate strategies [23, 24].

The first strategy is the crowdsourcing platform’s corporate strategy that provides crowdsourcing services to both business persons (orderers) and workers (contractors).

As mentioned earlier, the crowdsourcing platform understands the characteristics of the information society (necessary information can be used “anytime”, “anywhere”, “anyone”, “anyone”), and its business purpose and business plan. A business plan must be formulated based on (business design). And the crowdsourcing platform needs to develop its business by selecting its best position from the five positioning.

In other words, in the crowdsourcing platform, if the business design or business plan formulated by the company is wrong, it will handle non-proprietary fields (specialties), and both business persons (orderers) and workers (contractors). Therefore, the service content with poor quality will be provided.

As mentioned above, the crowdsourcing platform must formulate a business design by understanding the characteristics of the information society (necessary information can be used “anytime”, “anywhere”, “anyone”, “anyone”.).

And the crowdsourcing platform needs to develop its business by selecting its best position from the five positioning ((1) many items business type (general function type), (2) many items business type + high specialization business type (multifunctional type: advanced expertise type), (3) specific business type + high specialization business type (specific function type: advanced functions type), (4) many items business type + low specialization business type (multifunctional type: low expertise type), (5) specified work type + low specialty work type (specified functions type: low functions type)).

In other words, if the business design and business plan formulated by the company is misunderstood, the crowdsourcing platform will handle non-professional fields (specialties) and both business persons (orderers) and workers (contractors) Will provide poor quality service content.

In other words, in the crowdsourcing platform, if the business design or business plan formulated by the company is wrong, it will handle non-proprietary fields (specialties), and both business persons (orderers) and workers (contractors) therefore, the service content with poor quality will be provided.

The second strategy perspective is the business strategy of business persons (orderers) who use the crowdsourcing service provided by the crowdsourcing platform.

Business persons need to develop a business plan (business plan) based on their business purpose and business plan (business design), and then develop a business using crowdsourcing services.

Business persons understand the characteristics of the information society, and perform all operations within the company ("Sales", "Accounting", "Human Resources", "General Affairs", "Information System", "Development", "Research", "Production", "Public Relations", etc.). On the other hand, actively utilize the cheap labor force outside the company.

The third strategy perspective is a corporate strategy that corresponds to work style reforms for both business persons (orderers) and workers (contractors).

Japan has various labor environment issues such as "declining birthrate and aging population, decreasing labor force population due to low birth rate", "health adverse effects due to long working hours, and restrictions on working style due to childbirth, childcare, and nursing care".

In particular, gaps in the work environment and restrictions on work styles affect the quality of work and detract from the work of workers.

The working environment in Japan is longer and less productive than overseas. In this situation, Japan will fall into a negative spiral, where productivity will not increase despite long working hours.

And business persons use crowdsourcing services to establish a competitive corporate environment for rival companies.

The third strategy perspective is a corporate strategy that corresponds to work style reforms for both business persons (orderers) and workers (contractors) [25–30].

In order to improve the working environment, the Japanese government is advocating "work style reform". Japan's work style reforms are not implemented for either business persons (orderers) and workers (contractors).

Work style reform is an initiative that promotes "workers happiness and growth" and "improvement of productivity and creativity of business persons" together with business persons and workers. For both business persons and workers, one of the approaches to work style reform is the use of crowdsourcing services.

6. Conclusion

In today's information society, crowdsourcing attracts attention as a new business model that connects unknown business person (orderer) and an unspecified number of worker (contractor) through Internet websites.

Crowdsourcing seeks to reform traditional employment by providing work that utilizes ICT. In conventional Japanese human resources employment, workers were mainly employed as permanent employees who entered into employment contracts with companies; or as part-time, temporary, or seasonal workers. In each case, they were required to work at a designated work place and time.

However, with crowdsourcing, business persons generally carry out business operations by contracting with a large number of outside individuals. As long as there is an Internet environment, workers can now work anywhere and anytime; in urban areas, rural areas, sparsely populated areas, islands, overseas, etc.

In addition, the use of crowdsourcing has changed the lifestyles of workers. For example, new jobs are provided to residents in depopulated areas and islands, and to people who are raising children, pregnant women, students, and freeters (part-time workers), etc. It also provides a new style of working. Further, it is possible to create new jobs for people who have high skills but cannot make use of their skills in marriage retirement, or due to temporal circumstances.

Crowdsourcing creates new opportunities for professionals and non-professionals alike, and can offer inexpensive services that previously required high compensation. In sum, the spread of crowdsourcing has created new options for how individuals work and live. In the future, it will be possible to explore new ways of connecting business persons and workers through crowdsourcing.

Acknowledgements

This work was supported by Japan Society for the Promotion of Science (JSPS) Grants-in-Aid for Scientific Research (KAKENHI) (B) Grant Number 17H03327 and JSPS KAKENHI (B) Grant Number JP19H01532.

Author details

Tetsuro Saisho
Kokushikan University, Japan

*Address all correspondence to: tsaisho@kokushikan.ac.jp

References

[1] Saisho T. Corporate Strategy Changed by Matching Business. Hakutwo Shobo; 2016

[2] Ministry of Internal Affairs and Communications. Impact on Society Brought About by Rapid Evolution of ICT, "Information Communication White Paper 2014 Edition". Ministry of Internal Affairs and Communications; 2014. Available from: http://www.soumu.go.jp/johotsusintokei/whitepaper/ja/h26/pdf/n4100000.pdf [Accessed: August 20, 2019]

[3] Small and Medium Enterprise Agency Edition. Small and Medium Enterprise White Paper. Nikkei Printing; 2014

[4] Blattberg E. oDesk and Elance Merge, Forming One Giant Freelancer Company (Updated), "Entrepreneur". VentureBeat; 2013. Available from: http://venturebeat.com/2013/12/18/odesk-elance-merger/ [Accessed: August 20, 2019]

[5] Freelancers Union & Elance-oDesk. Freelancing in America: A National Survey of the New Workforce. 2014. Available from: https://fu-web-storage-prod.s3. amazonaws.com/content/filer_public/c2/06/c2065a8a-7f00-46db-915a-2122965df7d9/fu_freelancinginamericareport_v3-rgb.pdf [Accessed: August 20, 2019]

[6] Elance-oDesk. Freelancing in America: A National Survey of the New Workforce, LinkedIn. 2014. Available from: https://www.slideshare.net/oDesk/global-freelancer-surveyresearch-38467323 [Accessed: August 20, 2019]

[7] Mori Y. Cloud Computing-Technology Trends and Corporate Strategy, Ohmsha; 2009

[8] Koike R. Future of the Cloud-A World of Super Concentration and Super Dispersion; Kodansha; 2012

[9] Yoshida K. "The Business Changes with Crowdsourcing", Diamond. Ministry of Internal Affairs and Communications (2014), Impact on Society Brought about by Rapid Evolution of Ict, "Information Communication White Paper 2014 Edition", Ministry of Internal Affairs and Communications; 2014

[10] Crowdsourcing Japan. Crowdsourcing Five Classifications (crowdsourcing.org), Crowd Power Partners. 2013. Available from: https://crowdsourcing.jpn.com/2013/06/09a003/ [Accessed: July 20, 2019]

[11] Crowdsourcing Site Construction System. 2015. Available from: http://www.cs-system.com/ [Accessed: August 20, 2019]

[12] Atsushi Naito. Toward Crowdsourcing Specialization, Crowdsourcing from the Perspective of Foreign Players Now, TechCrunch Japan. 2014. Available from: http://jp.techcrunch.com/2014/04/07/crowdsourcing-trend/ [Accessed: August 20, 2019]

[13] Massolution. Crowdsourcing Industry Report: Enterprise Crowdsourcing: Market, Provider and Worker Trends, Crowdsourcing.org. 2012. Available from: http://www.crowdsourcing.org/document/enterprise-crowdsourcing-research-report-by-massolution-market-provider-and-worker-trends/13132 [Accessed: August 20, 2019]

[14] Planview Spigit. The 2018 State of Crowdsourced Innovation Report; Spigit. 2018. Available from: http://go.spigit.com/rs/123-ABC-801/images/2018_Spigit_State_of_Crowdsourced_Innovation.pdf [Accessed: August 20, 2019]

[15] Abrahamson S, Ryder P, Unterberg B. Crowdstorm: The Future

of Innovation, Ideas, and Problem Solving. Wiley; 2013

[16] Nishida R. Forty Percent of Cloudworks Projects that can be Hired by Engineers to be Used on an Hourly Basis, "CrunchBase". Japan: TechCrunch. 2012. Available from: http://jp.techcrunch.com/2012/07/18/jp20120718crowdworks-data// [Accessed: August 20, 2019]

[17] Yano Research Institute. Survey Results on Crowdsourcing Service Market 2014. Yano Research Institute; 2014

[18] Yano Research Institute. 2014 Crowdsourcing Market Reality and Prospects. Yano Research Institute; 2014

[19] CS Japan. Crowdsourcing Japan, Crowd Power Partners. 2019. Available from: http://crowdsourcing.jpn.com/ [Accessed August 20, 2019]

[20] Brandon KH. Crowdsourcing Services 15 Selections in the US, "Startup", btrax. 2012. Available from: http://blog.btrax.com/jp/2012/12/03/crowdsourcing/ [Accessed: August 20, 2019]

[21] Gillian T. Australia's Freelancer Soars on IPO Debut, The Wall Street Journal. Dow Jones & Company; 2013. Available from: http://blogs.wsj.com/moneybeat/2013/11/14/australias-freelancer-soars-on-ipo-debut/ [Accessed: August 20, 2019]

[22] CrowdNote. Crowdsourcing and Printing Mail Order Comparison List, CrowdNote. 2015. Available from: http://newcrowdnote.net// [Accessed: August 20, 2019]

[23] Kawaraban Editorial Department. Learn Crowdsourcing from American Cases, "Working Style", Zenken. 2015. Available from: http://w-kawara.jp/money/huge-black-ships-looming-quietly-2/ [Accessed: August 20, 2019]

[24] Junya M. Mechanism of 'Crowdsourcing' that Diversify Usage, "Meet Recruit", Recruit Holdings. 2014. Available from: http://www.recruit.jp/meet_recruit/2014/12/pr03.html/ [Accessed: August 20, 2019]

[25] Gratton L. The Shift: The Future of Work Is Already Here. Harpercollins; 2011

[26] Chikirin. Let's Think About How to Work in the Future-Life can be Lived Twice. Bungeishunju; 2013

[27] Sato A. New Work at Home: Introduction to Crowdsourcing, Impressions; 2014

[28] Gratton L, Scott A. The 100-Year Life: Living and Working in an Age of Longevity. Bloomsbury Information; 2016

[29] Okubo Y, Minazuki M. Work Style Reform: Management that Utilizes Individuals. Nihon Keizai Shimbun; 2017

[30] Akio Tsuchida, Deloitte Tohmatsu Consulting. Work Style Reform: Seven Designs. Nihon Keizai Shimbun; 2017

Chapter 10

Integrating Information and Communication Technology in Entrepreneurship in Sports: The Way Forward

Oliver Igwebuike Abbah and Uzoamaka Ogwo

Abstract

The topic delved into the use of Information and Communication Technology in sports entrepreneurship, in Nigeria. It discussed entrepreneurship in relation to small and medium scale sports enterprises. The topic explored the areas that Information and Communication Technology can be employed in sports entrepreneurship and the various types of ICT facilities that can be integrated into sports entrepreneurship at the small and medium scale level. Authors also discussed the challenges of integrating information and communication technology in sports entrepreneurship at this level. Finally, the way forward delved into strategies that can possibly be used to further the use of ICT in sports entrepreneurship in a developing economy like Nigeria.

Keywords: information and communication technology, entrepreneurship, sports, sports entrepreneurship

1. Introduction

The sports industry is growing extensively. It is a dynamic and unique industry that is inherently entrepreneurial on numerous fronts. Sports have progressed from a pure pursuit of leisure to a multi-million dollar industry [1]. It is a unique entertainment industry that has been growing in a geometric progression in a commercial environment. Sports provide a lucrative and continually growing marketplace worthy of immense investments [2], thereby providing not only economic impact, but also entertainment for millions of people globally. However, unlike other industries, sport has a unique combination of profit and non-profit organizations that work together to create a competitive environment [3]. While entrepreneurship literature has expanded to include various categories of entrepreneurship, including social, community based, and corporate entrepreneurship, few connections between sport and entrepreneurship exist in current literature.

Business organizations operate in a complex and competitive environment characterized by changing conditions and highly unpredictable economic climate. These changing environmental constraints not only affect their internal structure of business organizations but also their survival, growth and development. Thus, Information and Communication Technology (ICT) is at the centre of this global

IntechOpen

change. ICT directly affects how managers decide, how they plan and what products and services are produced. Sports entrepreneurs need to acquire specialized ICT skills in order to develop new and innovative way to satisfy the growing needs of sports consumers.

Reference [4] defined sports entrepreneurs as those persons who act as change agents in the supply of sports products, who attempt to increase the output of the industry, improve the consumer experience, or raise interest in sports products by such means as developing new markets and creating new products. Most discuss on sports entrepreneurs are directed to large scale enterprises by focusing on how new products emerge in the market and the role of ICT in sports innovation. Thus focus has not been directed towards small and medium scale sports entrepreneurship which a graduate of physical education and sports studies can delve into in other to be self-employed. This group of entrepreneurs needs to be developed especially in a developing country like Nigeria where graduate unemployment rate is steadily on the increase. This chapter therefore, focuses on the integration of ICT in the operations of small business enterprises in the area of sports.

2. Sports entrepreneurship in Nigeria

Sport generally defined, means any form of physical activity that aims to improve fitness or mental well-being. The emphasis is on physical expression but it can vary in intensity from low impact sports like walking to high impact sports such as aerobics. Hence [5] considers it to be organized physical activities through which exercise, amusement, fun and play are enjoyed. It provides a lucrative and continually growing market place worthy of immense investments that provides not only economic impact but also entertainment for millions of people globally [2]. The sports industry has thus developed into big business which individuals can harness for entrepreneurship ventures.

The concept of entrepreneur is changing as there are evolving ways to conduct business depending on the market context. An entrepreneur refers to a person, an initiator of business ventures who identifies and harness opportunities and bearing the risk to produce goods and services in order to satisfy the needs of the consumers to maximize profit [6]. Entrepreneurs are innovative individuals who have developed an on-going business activity to satisfy the existing needs of people in order to maximize profit. Entrepreneurs constantly discover new markets and try to figure out how to supply those markets efficiently and make a profit. He is a person that searches for change, respond to change, and exploits change by converting change into business opportunity, [7]. Characteristics of entrepreneurs include being self-motivated individuals who start enterprises relying on their initiative to satisfy existing needs [8]; having the ability to implement their vision and confidence in their ability to develop a business. This means that an entrepreneur is a leader and individual who identifies and solve existing problems. He is creative, innovative, resourceful, opportunist, risk-taker, initiator, skilled and open minded with vision and mission to motivate and drive a business venture with sense of value, team-building and managerial skills, [9].

Entrepreneurship development is the pivot point for economic growth globally. Reference [10] defines entrepreneurship as any attempt at new businesses or venture creation, such as self-employment, a new business organization or the expansion of an existing business by an individual, team of individuals or established business. It is the practice of starting new organizations or revitalizing mature organizations particularly, new businesses in response to identified opportunities [11]. In order words, entrepreneurship requires skill acquisition, critical thinking to know business opportunities, planning to map out strategies to execute them, and creative and innovative skills to develop new products [6]. These can be found in all fields of endeavor including sports.

The sports industry is growing extensively and offers variety of entrepreneurial opportunities. Sports entrepreneurship is the marketing of sports products and services to satisfy the sports needs of individuals and for profit making [6]. It entails the identification of conditions and procedures in which emergent business ventures with a social orientation are formed. In other words, it provides opportunities to promote innovations, investments, competitiveness, sustenance of sports business ventures and ability to create a market niche for income.

Sports entrepreneurship is a panacea for economic transformation in Nigeria as it encourages innovative activities that facilitate sustainable economic development. [5] noted that Sports organization in Nigeria has resources at their disposal that can be turned around for entrepreneurship development through sports marketing. [5] further noted that Sports also provide tremendous business opportunity in merchandising, whole lots of manufacturing industries have emerged producing sports attires, sports equipment, sports sourvenirs, and services for sportsmen, women and fans. These sports industries are contributing immensely to entrepreneurship development in Nigeria. Sports entrepreneurship is supposed to thrive in Nigeria considering its population and the passion Nigerians have for sports. With a population of over 180 million of which 65% are between 15 and 35 years [12], Nigeria guarantees a long term market for sports entrepreneurship. This is complemented by a TV and media audience of between 25 and 50 million- capable of attracting a fair share of the over N400 Billion advertising spent on sports content and events; and a mobile and internet growth projection of 10% annually [12]. The creation and development of small and medium scale enterprises in the sports ancillary sectors such as merchandising, equipment manufacture as well as investment in venues and facilities construction and content development, management and broadcast, is capable of adding considerably to the size of the industry [12]. However, [13] lamented that the Nigerian sports landscape is replete with immense, yet untapped, business opportunities. The development of entrepreneurial practice among sports experts (especially graduates of Human Kinetics and sports) will go a long way to reducing graduate unemployment rampant in Nigeria.

Sports entrepreneurship has not really been harnessed by Physical education graduates in Nigeria. An industry as large as the sports industry requires educated people to run a variety of sports related businesses [14]. The sports industry entails a variety of entrepreneurial outfits. These may include: health club facilities, sports arena and facility operators, league owner/operators, sporting goods store owners, sports ticket agencies, and sport physical therapists – just to name a few [14]. [14] further stated that there has been significant growth in the coaching and fitness type businesses in recent years within the sports industry. In a study by [15] 23.19 per cent of Physical Education students undertook employment in sports-related occupations including sports coaching in a variety of sports, fitness gym coach, dance instructor, life guard and personal trainer. [16] equally reported that the most entrepreneurial engagement of graduates of physical education is sports events centres. Entrepreneurial practices of Physical education graduates include organizing sports events like inter-house, inter-community sports, etc. (21.30%), coaching (19.03%), establishment of health and fitness club (11.60%) and dance instructor (10.97%) [17]. In other words, organizing sports events like inter-house, inter-community sports, etc. is the most entrepreneurial practice engaged in by Physical education graduates, because most of them are in the teaching profession. This is partly due to the curricular deficiencies of the Physical education programme in Nigeria. The nature of the curriculum utilized for the preparation of prospective physical education specialists lacks entrepreneurial skills. This is so because the curriculum as stipulated by the National Universities Commission (NUC) benchmarks and minimum standards consistently emphasized school-based Physical Education solely aimed at training prospective graduates to become teachers of Physical Education in schools. This has adversely affected the profession by restricting the role it could play in developing entrepreneurial skills and creating newer job opportunities in order to enhance national economic growth.

Possible entrepreneurial areas in sports that has been neglected by vast majority of physical education graduates include such businesses like sports magazines and newspaper, sports viewing centres, owning a sports shop, running a sports television and radio show, becoming personal trainers, coaching, owning an academy and a gymn [18]. Others include becoming a football scout, nutrition coach, owning a local sports club, sports blog, public relations outfit for athletes, sports betting company and a sports jersey customization business.

Physical education graduates if well exposed to entrepreneurial skills can start a sports magazine or newspaper. Most people in Nigeria are not yet internet connected and as such still depend on the print media for news including sports news. This explains why a lot of people still flock newspaper stands today on a daily basis in order to read the latest news. Most sports lovers are not exempted from this practice as they follow news about their favorite players, athletes and football clubs both at home and abroad. A typical example of such magazines was SportsWorld floated in the 80s in Nigeria and also Complete Sports.

Sports viewing centres are also sports entrepreneurial outfit open to physical education graduates. Due the passion Nigerians have about sports, sports viewing centres are well patronized. During tournaments, tennis match, national and international football matches, club matches, etc.; such viewing centres are always filled up. This is because fans prefer where they will interact with fellow fans and engage in sports discussions about their clubs and favorite players. Most of these centres in Nigeria are floated by non-professionals.

Opening sports shop for sale of sports equipment and supplies are also sports entrepreneurial options that can be lucrative. So many people in Nigeria today engage in exercise and fitness workouts. Such individuals need sports wears in other to participate actively in such programmes. This can include producing customized jerseys and sportswear for individuals and teams. It is also not uncommon to find people wear sports wears for non-sporting activities and gathering either for the fun of it or to show solidarity and support for a team or sport among the people. Moreover, with the ever increasing population of the country, sports shop outlets are promising business. By starting a sports shop in a well visited sports environment, you can serve the needs of fitness junkies by selling sport shoes, trophies, gym supplements, gloves, jerseys, face towels, shakers, and a lot more [12].

Becoming a personal trainer can equally be a lucrative business venture. With the current rise in over weight and obesity brought about by changes in lifestyle and nutritional practices, a great number of people want to train in order to reduce or watch their shape. A lot of women for example, want to look beautiful after childbirth. Such women require the services of well trained professionals to guide and prescribe exercise for them. With a large market looking for ways to shed weight, [12] stated that one could apply for the job of a fitness trainer at a hotel, health centre, and more, to train people daily. More so, some of the trainees may request private sessions from their homes.

Sports coaching have become a very lucrative business in the world today. This is an entrepreneurial option that graduates needs to explore because so many athletes needs to start their career early in order to acquire the basic fundamentals and necessary complex skills in their chosen sport. The key element is for them to undergo the right training to be able to improve their skills to a national or international level. This ability to a great extent depends partly on the ability of a good coach to guide them. Areas of specialization abound in the field of sports coaching. These include association football (soccer), athletic, basketball, volleyball, hockey, badminton, tennis, table tennis, squash, handball, squash, etc.

Establishment of sports academy gives young sportsmen and women the opportunity to hone their talents, and also offers them exposures that increase their chances of being spotted by higher clubs. This has over time proven to be a profitable venture. By starting a sports academy that trains would-be professional footballers, you will not just be responsible for the success of many future players, but would build a profitable business in the process [12].

Establishing or working in a multi-gym shop is one area with the potential of accommodating a lot of graduates in the area of sports entrepreneurial outfit. Many people flock the gyms and fitness centres these days just to maintain a sexy look. This can be attributed to the strong influence of social media. With the increasing number of people looking for places to engage in a workout, setting up a standard and affordable gym in a residential area is a smart sports business startup.

Starting a sports blog can also be an entrepreneurial outfit worthy to be ventured into. By taking advantage of people's cravings to find solutions to various health or fitness problems, starting a sports blog focused on fitness and general wellbeing could be a great bet [12].

Blogs can serve various purposes such as telling sports stories through a personal blog or media channel. One could also run a fitness blog, where people could learn various ways to stay healthy, how to work out right, and what meals to consume to be in top shape.

Sports betting have become one the most lucrative business enterprise today. So many people are addicted to placing bets on various matches every time there is a game to be played. According to [12], the average revenue generated from sports betting companies daily in Nigeria, is about $7 million dollars. This has made the government to hike the cost of obtaining sports betting license to fifty million naira as at 2016.

3. Areas that ICT can be employed in sports entrepreneurship

In the present digital environment, Information and Communication Technology (ICT) is central and a key factor to economic and national development. It is a pivot point around which every economic and social development revolves globally. Information and Communication Technologies (ICTs) are playing a pivotal role in every walk of life. Information and communication technology (ICT) emerged from the fusion of information technology (IT) and communication technology. In explaining the concept of ICT, [19] defined (ICT) as the handling and processing of information using electronic devices. ICT is defined as the creation, collection, storage, processing, transmission, display and use of information by people and machines [20]. ICT is defined as the various technologies that facilitate communication, processing and transmission of information, goods and services by electronic means. It consists of hardware, software, networks and media for the collection, storage, processing, dissemination and presentation of information (such as audio, visual, text, images) as well as related services. Both traditional technologies (telephone, fax, radio, television, and print media) and newer technologies (such as mobile phones, computers, and internets) are usually included in the concept of ICT [21, 22].

Information and communications technology (ICT) describe the variety of technological tools and resources used to produce, distribute, store and manage information and knowledge [23]. These technological tools include all kind of computer system, software and applications and networking systems which facilitates communicate within the digital world, [24]. To [25], information and communication technology offers new innovation modes for all students at all education levels. In other words, they are those technological tools and resources utilized to create, store, manage and disseminate information. [26] identified these tools and resources as computers, the internet, broadcasting technologies such as radio and television and telephone (including mobile phones). ICT is at the centre of a global change curve that have directly affected how managers decide, how they plan and what products and services are produced. ICT has dramatically transformed the lives of individuals. It provides businesses the scope to analyze data and plan business strategies accordingly. Utilizing ICT means that the data analysis is accurate, thus optimizing profits. ICT has permeated all aspects of human endeavor globally.

The integration of these technologies in sports has brought a paradigm shift in sports industry and has dramatically improved the operations, products and services delivery in sports industry. ICTs have brought about a lot of innovations and revolutionary transformations in sport industry. In cricket, hot spot technology is used to gain information and make precise decision. In football goal line technology is used to detect when the ball crosses the goal line. In sports like badminton and lawn tennis, hawk-eye technology is used to make correct decision [24]. The application of cutting-edge Information and Communication Technologies in Sports has the potential to boost Nigerian Economy and entrepreneurial opportunities for the teeming population. It has a lot of potential to ensure efficiency in sports database management, enhance collaboration, co-ordination, global visibility, easy access to information and networking in sports industry for sustainable economic and social development in Nigeria. ICT will in no small measure enhance productivity of sports products, services, advertisement and marketing. On this note, [27] agreed that all sectors of the economy (trading, manufacturing, services, culture, entertainment, education, medical, transportation etc.) have a lot to benefit from the existing information and communication technologies such as micro-controllers, personal computers, internet access, mobile phones, digital video conferencing, e-mail, multimedia among others. This has offered most Nigerian youth employment opportunities by establishing outlets of sports products and services where people watch football match and other sports activities as well as predict sports outcomes, for example *BetNiger*. The field of sports entrepreneurship has certainly been affected by the penetrating influence of ICT the world over particularly in developed countries. ICT has resulted in very profound and remarkable changes on the quality and quantity of business that can be transacted in the area of sports. Accordingly, ICT applications in sports entrepreneurship has the potentials to accelerate, enrich, deepen and motivate business in the area of sports.

ICT can play a major role in the procurement and supply of sports products and services. This can be achieved through accessing electronic procurement systems. Electronic procurement is an internet-based business process for obtaining materials and services, and managing their inflow into the organization [28]. This is a business-to-business or business-to consumer purchase and sale of supplies and services through the internet as well as other information and networking systems. Electronic procurements includes the use of purchasing cards, reverse auctions, integrated automatic procurement systems to facilitate the buying process. This helps in providing global visibility of sports products and services to allow sports manufacturing companies to have better access and control over their products as well as information flow across supply chain.

Electronic procurement can help to reduce problems associated with procurement and supplies, track data and make it easier for sports entrepreneurs to make their purchases and track their progress. E-Procurement, also known as electronic procurement or supplier exchange, is the purchase and sale of supplies, equipment, works and services through a web interface or other networked system. The technology is designed to centralize and automate interactions between an organization, customers, and other value chain partners to improve speed and efficiency of procurement practices [29]. Typically, E-Procurement web sites can be used to look for buyers or sellers of sports goods and services. Interested buyers and sellers may specify costs of the sports good. E-procurement can also reduce purchasing agent overhead costs. E-Procurement activities that can be utilized by sports entrepreneurs include online indents and demand aggregation, online expression of interest, online bid submission, system supported evaluation, online status publishing, online release of purchase order, letter of award, and order fulfillment and post procurement processes.

ICT provides the best tools for communicating with producers, wholesalers, customers and other business partners. It has unlocked the facilities like e-mail, social media and other messaging platforms for purposes of communication. This can be accessed for making inquiry and contacts by sports entrepreneurs. Also, information search is best achieved through ICT facilities. ICT can positively affect information processes and the information made available to sports entrepreneurs in making business decision.

ICT can be used by sports entrepreneurs in the design and manufacture of sports equipment and supplies. Sports equipment manufacturers can employ computer aided design such as 2D design, Pro-Desktop and Google Sketch up. These are computer software that allows the user to design their products on the computer. Others include rapid prototyping, virtual reality modeling and computer integrated manufacture. This can help to reduce the waste of materials as instant changes and developments can be sent virtually for partners to work on and make adjustments.

ICT can also play major roles in digital marketing of sports equipment and supplies. Digital Marketing according to the [30] is a term that encompasses modern marketing techniques used online, for example, Social Media, Blogging, Search Engine Optimization (SEO), Pay per Click Management (PPC), Branding, Content Marketing, Video Marketing and App creation. [31] stressed that nowadays, digital marketing is indispensable for successful sports retailers. If used properly it is much more than an ideal supplement to traditional advertising. With clever online marketing you can support product sales effectively, maintain your own image and build up your shop as a brand. Digital marketing of sports products and services is a key driver to economic development and is rapidly changing the way sports products are packaged, disseminated, accessed, consumed and sold to sports consumers. According to [32], sport has a unique marketing approach, as most of its messages and images are conveyed through media. ICT in marketing avoids storage costs as orders are readily available through quick response marketing. Also most products for sale are marked with a bar code. These bar codes are scanned at an Electronic Point of Sale machine. This means that individual items can be tracked from end of manufacture, to distribution, to shop and finally to sale. This is essential for JIT production and QRM. This implies that products do not need to be stored as only the essential amount of products are released at a time based on the stock currently held for example by the store.

Advertisement of sports equipment and services is another area that ICT can be employed by sports entrepreneurs. By advertising we refer to the promotion of an enterprise's products and services with the main aim of driving up sales products and services. The use of websites and social networking sites for the advertisement of sporting events have ushered in a modern and innovative means of disseminating sports news to a wider population. They play a key role in advertising sports events, goods and services. This can be harnessed by sports entrepreneurs to boost their sales and global visibility. Advertisements can be engaged in order to increase sales of product and service, create and maintain a brand identity or image, communicate changes in the existing product line, and introduce new product or service. The ubiquitous availability, access and use of smartphone technologies have continued to promote online branding and mobile internet marketing of sports events, sports goods and services. Therefore, ICT has the capability of transforming sports business if well harnessed.

ICT plays significant role in the area of business management. ICT systems can enable sports entrepreneurs to store, process, analyze and share vast amounts of data. The information available enables sports business managers and by extension the employees to make fast business decisions. Technology has over the years caused an explosion in business and commerce. Many sports enterprises can be revolutionized and remodeled to best practices due to the introduction of ICT in its management. Some of the areas in which technology is crucial to sports business include point of sales systems, the use of ICT in management, accounting systems, and other aspects of day to day running of sports business activities.

ICT can also be utilized for business record keeping purposes. The need for a records management in sports business enterprises cannot be overstressed in the digital age. This has to do with the creation, storage, accessibility and security of digital information. Daily sales and purchases records can be kept with the aid of ICT tools. The ultimate aim of both record management to support, protect and enable the sports business in a cost-effective manner. It is also important it has superior data storage capacity, faster data retrieval, quicker data sorting and analysis. The application of ICT the management of records in sports enterprises therefore, will go a long way in making such records accessible and usable.

Monetary transaction is another area ICT can be utilized in sports entrepreneurship. ICT has made possible various monetary transaction activities like electronic banking. This is simply the use electronic and telecommunications network for banking transactions (making payments for goods and services). Sports enterprises can have Point of Sale (POS) services, online transfers, and the use of credit and debit cards in their business outfits. This encourages non cash payments for products and services.

Storage and data backup is another area that ICT can be utilized in sports entrepreneurship. This can be carried out by use of cloud computing. Cloud storage is a service where data is remotely maintained, managed, and backed up over a network. This is also called storage virtualization. The service is available to users over a network, which is usually the internet. It allows the user to store files online so that the user can access them from any location via the internet. Storage virtualization could be a private storage, like those hosted by a company, or a public storage, like those hosted outside of a company.

ICT has been proven to enhance customer satisfaction in entrepreneurial outfit. The perceived effect of ICT on service quality and customer satisfaction was investigated by [33]. The study established that as the ICT service delivery increases, so does the customer satisfaction. To enhance customer satisfaction, there is need to increase the use of ICT in service delivery. The ICT service delivery affects customer satisfaction. [34] also reported that ICT positively affects customer satisfaction and the ease of carrying out business. In addition, ICT have created opportunities for sports equipment entrepreneurs to advertise their business online. This allows wide visibility and remote access for the product [6]. Customers can thus order for sports equipment and services online. They can equally make online payments without the barrier of physical boundary. In the area of sports equipment design, technology is also playing a growing impact [6]. [35] reported that the application of ICT to sports serves a role in creating whole new opportunities for entrepreneurs and provides sports events and tools such as Computer Assisted Design (CAD) which can play a very important role in sports equipment.

4. Types of ICT facilities that can be integrated into sports entrepreneurship

The rise of ICT has paved the way for various innovations. With the digitization of business through ICT, more and more businesses are increasingly leveraging the benefits of digital tools to improve their prospects and sports entrepreneurs are not exceptions. Information technology infrastructure includes computer hardware, software, data, storage technology, and networks providing a portfolio of shared information technology resources for the enterprise [36]. ICT facilities that can be integrated into sports entrepreneurship can be categorized into two: software and hardware.

Hardware refers to any physical part of the computer system which you can physically touch, hold, pick or move. Software on the other hand, refers to a set of instructions which tells the computer what to do. These can be integrated for one purpose or the other in a sports enterprise.

4.1 Computers

Computers became so popular in business establishments solely for secretarial purposes namely word processing and spreadsheet management. However, computers are today, used to carry out different functions in a business enterprise. Computers that can be used in business include desk tops, laptops, notepads, servers and even smart phones. Their uses in business are endless. Different kinds of programs and operating information, known as software, are used by computers to do specific tasks. Productivity tools such as Microsoft Word, a word processing package, and Microsoft Excel, a financial spreadsheet system are used software used by business entrepreneurs. Businessmen/women choose software depending on what they intend to do.

Computers are now a vital tool for communication. Communication is the soul of any business enterprise especially in sports. It is the life wire and key driver of sustainable sports development around the world. Computers have enabled concepts such as remote working and flexible working schedules. It has also enabled businessmen and women to communicate with their customers. This is achieved through e-mail, Skype, zoom, short message system, social media platforms, etc. It has made it easier for customers to communicate with business enterprises in a timely and efficient manner. The biggest change to the sports industry has been the advent of social media. With platforms such as Twitter, Facebook, Blogging, Instagram, YouTube etc. There are millions of users of sports brands that sports entrepreneurs can use to promote their sports products and services, [30]. The use of ICT in sports has bridged the gap of physical boundary as sports fans can view sports activities from any part of the world.

Computers are also used for marketing of products and services. This is possible when the business outfit has internet connectivity [37]. The entrepreneur can also utilize different social media platforms to market their products and services. Computers are also very important as an accounting tool. They are critical for performing such tasks as preparing invoice for customers, calculating payroll, keeping income and expenditure accounts, among others. There is online accounting software like 'Xero' which does the work of an accountant and book keeper. It has the ability to track expenses, profits, and reconcile debts. It can also send invoices and create expense claims.

4.2 Digital video camera

The use of digital video cameras plays an indispensable role in recording performance of players in the field. The recorded video can be broadcasted live or uploaded later on. It can be used to record motion pictures, capture moving images and synchronize sound. Recent advanced technology used high-tech video recorder to record any performance and can directly convert them to three-dimensional data using high level software, [24].

4.3 Printers

Printers are necessary in the event that you have the need to print out your marketing materials or need to handle basic jobs. Printers will enable you minimize costs by encouraging your employees to work as a paperless office whenever possible [38]. Printers are also needed for printing of contracts, legal documents, and other business materials. Some printers have multiple functions including scanning, photocopying, and faxing [38].

4.4 Data storage devices

External Hard Drives are used to store certain types of information and help you stay organized. It is an important tools for entrepreneurs especially those that do not want to store information in the Cloud or on their computers' hard drives. External hard drives have the ability to protect valuable and sensitive business data in the event of an accident, such as a fire. It can also be used as back-up devices.

4.5 Cloud computing

Cloud computing has offered a paradigm shift in sports industry and it is currently gaining grounds, popularity and acceptance globally. Cloud computing holds a lot of potentials, innovations and the capacity to transform the sports industry to a multimillion dollar business enterprise. Cloud computing involve storing data and information away from our physical location as against storing them on local hard disk or local servers. The term cloud computing means storing and accessing data and programmes over the internet from a remote location or computer instead of in the computer hard drive [39]. This is as opposed to local storage and computing (storing data or running a prgogramme from the local computer hard drive). Therefore, for it to be considered *cloud computing*, one need to access data or programmes over the internet. The end result is the same; however, with an online connection, cloud computing can be done anywhere, anytime, and by any device such as smartphones, tablets, laptops that is connected to the internet. When the need to use such data or information arises, it is accessed and obtained through the internet. This makes such data and information accessible from anywhere. This is the essence of cloud computing.

The concept of cloud computing is immensely popular among entrepreneurial outfits due to the efficiency it provides in business operations. Some of the noteworthy benefits are cost savings, remote working, efficiency, flexibility, future proofing, morale boosting, and resilience without redundancy. It offers speedy, on-demand and easy access to sports data, products and services regardless of time and geographical boundaries. Adequate security of sports data from natural disaster, power and server failure is ensured. Cloud computing provides improved agility and time and resource management for businesses. Businesses are increasingly shifting to the clouds to leverage its many benefits. For cloud computing to be effectively utilized, a good internet connection and a commendable WiFi connectivity are important.

4.6 Mobile phone

Mobile phone can be used to a great extent for short message service and multimedia message service. This can be used for communication purposes and advertisement placements. With over a billion handsets worldwide with majority being internet enabled, mobile phones have been one of the fastest adopted consumer products of all time which can be harnessed for by entrepreneurs. Mobile phone has made possible wireless applications that enable person-to-person messaging, email, banking, news, games, music, shopping, ticketing, and information feeds. A lot of information and even advertisement can be carried out in Facebook, WhattsApp, You tube, etc. These applications can be integrated into sports entrepreneurship in order to enhance business activities. [40] noted that Sport can provide meaningful content for mobile technology, which entails the use of cellular communication technology to distribute information to consumers via cellular phones.

4.7 The internet

The internet is a global communication network that provides direct connectivity to anyone over a local area network. The internet service can be accessed through an internet service provider (ISP). It is a public network that is connected and routed over gateways. The ISPs are connected to the network service providers, and eventually to the internet appliances. Communications has been enhanced to a great extent through the internet. The internet has enabled improved internal communication, improved business partnership channels, effective marketing, sales, and customer support and facilitated collaborative activities support. In the area of business, the internet has enabled faster time to market goods and services, potential for simultaneous engineering and collaboration among business enterprises, lower design and production costs of goods and services, improved customer relationships, and creation of new business opportunities. It has also reduced operational costs for business enterprises. These can to a large extent improve the business outfits of sports entrepreneurs if well harnessed.

4.8 Business process automation (BPA)

ICT has enabled the automation of business processes. Business process automation (BPA) is the automation of the processes of business through technology. This allows businesses to cut costs and increase productivity. It also improves efficiency, increases workflow and saves time. BPA basically implies automating processes and workflow such as document approval process, employee on-boarding process, billing, tracking metrics, collecting customer data, monitoring, etc. Automation software is available for such purposes.

4.9 Network server solution

Network server solution is important if the sports entrepreneur will need any type of network to support the data base, email applications, and other files in the business. The entrepreneur has the option of choosing from several server solutions based on the amount of storage needed in the business, security requirements, and backup options required. There is need for the entrepreneur to work with a specialist in the area of networking to ensure the right package based on current and future needs are procured.

4.10 Wireless router

Wireless router is important to ensure that the business outfit is connected to the internet wirelessly. A strong wireless connection can ensure high-speed connectivity to all linked computers and mobile devices [38]. It will also reduce the need for extra cables running in and around the business enterprise.

5. Challenges of integrating ICT in sports entrepreneurship in Nigeria

Notwithstanding the potentials of ICT in sports industry in Nigeria, there have been a lot of factors militating against the effective integration of ICT in sports entrepreneurship in Nigeria such as the following:

- **Lack of ICT and e-business skills:** lack of fundamental ICT skills such as navigating the internet and troubleshooting is a serious deterrent for most people who intend to venture into sports entrepreneurship in most developing countries. There is lack of ICT technical expertise needed in managing sports businesses and this is one of the serious barriers in developing countries

- **Lack of Locally Manufacturing Industry**: most ICT facilities used for sports in Nigeria are imported products and there is a very high import duty on such facilities which make them very costly and unaffordable. This has been a bottleneck for sports entrepreneurs who may not have enough capital.

- **Poor Internet Connectivity:** poor network and broadband internet connection has been a bottleneck for the successful implementation of ICT in sports entrepreneurship in Nigeria. The growing digital divide in developing countries is a serious impediment to the utilization of ICT in sports entrepreneurship.

- **Lack of awareness:** ICT application in sports entrepreneurship has the potential of creating millions of job opportunities which are yet untapped. Most people in Nigeria are not yet exposed to the potentials of ICT in sports entrepreneurship.

- **Cost:** the initial cost of purchasing the ICT facilities such as computers, smart phones, internet connection, bandwidth, hardware, software and maintenance cost has been a very serious challenge facing potential entrepreneurs who want to venture into sports business.

- **Lack of Government incentive for sports entrepreneurs**: there is lack of incentive by the Nigerian Government to support sports entrepreneurship. The sports betting sector is blossoming and growing astronomically in Nigeria. However, there is lack of incentive for young Nigerian entrepreneurs who have the passion to venture into such entrepreneurial opportunity. Hence, they are faced with the challenge of acquiring the ICT facilities such as mobile phones and internet connection required for such sports business venture. Some young sports entrepreneurs who have the passion to establish sports viewing centres do not have the capital.

- **Epileptic Power Supply**: one of the major constraints for integrating ICT in sports entrepreneurship in Nigeria is poor power supply. Most sports entrepreneurs usually run the business centres on power generators, adding to cost of sports goods and services. This has been a very serious setback to the integration of ICT facilities in sports entrepreneurship in Nigeria.

6. The way forward in the use of ICT in sports entrepreneurship in Nigeria

The success of integrating ICTs in sports entrepreneurship hinges on proper training, adequate funding, motivation and creating enabling environment for potential sports entrepreneurs to encourage best practices in sports industry in Nigeria. The current transformation, innovation and revolutionary changes brought by the integration of ICT in sports industry demands an urgent revamping and repositioning of sports industry in Nigeria. In order to ensure the integration and sustainability of ICT in sports industry to boost entrepreneurship in Nigeria, the following measures should be adhered to:

- Nigerian government need to ensure adequate human capacity training on ICT skill acquisition to enable individuals develop innovative ICT skill, knowledge and capacity to boost sports entrepreneurship in a competitive global digital environment. There is need for government, policymakers and stakeholders in sports industry to ensure the integration of ICT and e-business skills acquisition into education curriculum at all level. Such training and skill acquisition must be proactive, tangible, and in real time where all the necessary tools, machines, equipment and ICT facilities should be adequately provided.

- Creating an enabling environment by the government to encourage investments in local content production especially in the area of sports goods and services using ICT facilities.

- Access to broadband internet connectivity has the potential of global visibility, connecting people, places, businesses and services without the problem of physical boundary. Sports business owners highly depend on broadband internet connectivity for global access, easy marketing and effective distribution of sports goods, products and services. Hence, government should ensure the availability and accessibility of internet connection is very critical for the effective implementation of ICT is sports entrepreneurship in Nigeria

- Creating awareness through sensitization programmes on the potentials of integrating ICTs in sports industry in Nigeria to boost sports entrepreneurship

- Government should subsidize the cost of ICT facilities to encourage potential entrepreneurs who may want to venture into sports businesses. There is need to give incentives to young sports entrepreneurs to start off their business

- Tackling the issue of poor power supply in Nigeria to encourage easy business operation is very critical.

Author details

Oliver Igwebuike Abbah[1*] and Uzoamaka Ogwo[2]

1 Department of Human Kinetics and Health Education, University of Nigeria, Nsukka, Nigeria

2 National Centre for Energy Research and Development, University of Nigeria, Nsukka, Nigeria

*Address all correspondence to: oliver.abbah@unn.edu.ng

References

[1] Chadwick S. From outside lane to inside track: Sport management research in the twenty-first century. Management Decision. 2009;**47**(1):191-203

[2] Research P. Cited in D. Ciletti (nd). In: Sports Entrepreneurship: A Theoretical Approach... FiT... Available Online. 2009 http://fitpublishing.com/sites/default/files/excerpt_sportsentrepreneurship.pdf

[3] Ratten V. Entrepreneurship in sport policy: A new approach. In: Ratten V, editor. Sport Entrepreneurship and Public Policy: Building a New Approach to Policy-Making for Sport. 2020 Springer Nature Switzerland AG

[4] Porter D, Vamplew W. Entrepreneurship, sport, and history: An overview. The International Journal of the History of Sports. 2018;**35**(7-8):626-640 https://www.tandfonline.com/doi/full/10.1080/09523367.2018.1544126

[5] Eze, W. U. (2015). The implications of sports and sports marketing for Entrepreneurship development in Nigeria. *International Journal of Current Research and Academic Review, 3(6):204-210. http://www.ijcrar.com/vol-3-6/Eze%20,Willy%20U..pdf*

[6] Uzoamaka O, Abbah OI, Igwe NJ. Perceived impact of information and communication Technology in Enhancing Entrepreneurship in sports among health and physical education teachers in Anambra state, Nigeria. In: (2019). *Library Philosophy and Practice (e-journal)*. 2712 https://digitalcommons.unl.edu/libphilprac/2712

[7] Arogundade BB. Entrepreneurship education: An imperative for sustainable development in Nigeria. Journal of Emerging Trends in Educational Research and Policy Study (JETERAPS). 2011;**2**(1):26-29

[8] Mueller SL, Thomas AS. Culture and entrepreneurial potential: A nine country study of locus on control and innovativeness. Journal of Business Venturing. 2000;**16**:51-75

[9] Ogwo, U., Ayolugbe, C. I. and Igwe, N. J. (2019). Entrepreneurship Education: A Viable Tool in Enhancing Post Basic in Awka Education Zone, Anambra State, Nigeria. *Library Philosophy and Practice (e-journal)*. 2713. https://digitalcommons.unl.edu/libphilprac/2713

[10] Harding R. Entrepreneurs: The world's lifeline. Business Strategy Review. 2006;**17**(4):4-7

[11] Onuoha G. Entrepreneurship. AIST International Journal. 2007;*10*:20-32

[12] Obi, S. (2018). The Sports Business – Challenges and Prospects. Downloaded online from: http://www.thesparkng.com/the-sports-business-challenges-and-prospects/

[13] The Lagos Business School (2020). Sports Business Management Programme. https://www.lbs.edu.ng/open_seminars/sports-business-management-programme/

[14] Borgese A. Educating sports entrepreneurs: Matching theory to practice. The Sports Journal. 2010:1-4

[15] Jones P, Jones A. (2014). Attitudes of sports development and sports management undergraduate students towards entrepreneurship. A university perspective towards best practice. Education and Training Vol. 56 No. 2014;**8**(/9):716-732

[16] Ogbonna RS. Entrepreneurship Engagements of Physical Education Graduates in Rivers. Nigeria: Unpublished Bachelors Degree Project, University of Nigeria, Nsukka; 2015

[17] Abbah OI, Nji GC, Badaru JU, Ekong IE. Entrepreneurial practices of physical education graduates in Abia state, Nigeria. International Journal of Studies in Education. 2020;**16**(1):41-53

[18] Edom, S. (2016). 15 Lucrative Sports Business Ideas & Opportunities. Available online from: https://startuptipsdaily.com/sports-business-ideas/

[19] The American National Council for Education (ANCET, 2000) Cited in Adegbija, M. V. (2003). Information and communication technology in women education in the new millenium". Forty Years of Educational Technology in Nigeria, 2, 16-24.

[20] Open University. Information technology living with technology: Block 2. In: Foundation Course Milton the Open University. 2006

[21] DFID (2002).The significance of Information and Communication Technologies for reducing poverty. In: Derbyshire, H. (2003). Gender issue in the use of computers in education in Africa.

[22] UNESCO, (2004).The significance of Information and Communication Technologies for reducing poverty. In: Derbyshire, H. (2003). Gender issue in the use of computers in education in Africa.

[23] Majoka MI, Fazal S, Khan MS. Implementation of information and communication technologies (ICTs) in education course: A case from teacher education institutions in Pakistan. Bulletin of Education and Research. 2013;**35**(2):37-53

[24] Gogoi H. The Use of ICT in Sports and Physical Education. Advances in Physical Education and Sports Sciences. Publisher: AkiNik Publications; 2019. pp. 117-130 https://www.researchgate.net/publication/337366737_The_use_of_ICT_in_Sports_and_Physical_Education

[25] Anujeonye, N.C.(2007). Information and communication technology (ICT) facilities in functional computer education for economic empowerment and development.*A paper presented at the 2007 annual school of science conference held at the federal college of education*, *Zaria* from 2nd–5th April.

[26] Udezi SO. The role of ICT in education. Nigerian Journal of science and educational research. 2008;*4*:55-61

[27] Taiwo TO, Adeyanju IA. The role and Prospect of information and communication Technology in National Development. International Journal of Computing Academic Research (IJCAR). 2013;**2**(3):113-119 Retrieved from: https://silo.tips/download/the-role-and-prospect-of-information-and-communication-technology-in-national-de

[28] Kameshwaran S, Narahari Y, Rosa CH, Kulkarni DM, Tew JD. Multiattribute electronic procurement using goal programming. European Journal of Operational Research. 2007;**179**(2):518-536

[29] Wax Digital (2009). What is eProcurement? https://www.waxdigital.com/glossary/what-is-eprocurement/

[30] Forceful Digital Marketing Agency (2020). Paradigm-shift in sports industry with digital marketing. https://www.bidigital.co.uk/blog/digital-marketing/paradigm-shift-in-sport-industry-with-digital-marketing/

[31] Spies, A. (2016). Digital marketing: Tips for sports retailers. https://www.ispo.com/en/know-how/id_77016932/digital-marketing-tips-for-sports-retailers.html

[32] Bruce T, Tini T. Unique crisis response strategies in sports public relationships: Rugby league and the case for diversion. Public Relation Review. 2008;**34**:108-115

[33] Otiso KN, Chelangat D, Bonuke RN. Improving the quality of customer service through ICT use in the Kenya power and lighting company. Journal of Emerging Trends in Economics and Management (JETREMS). 2012;**3**(5):461-466

[34] Kabanda G. The impact of ICTs on customer service excellence in Zimbabwe. International Journal of Emerging Technology and Advanced Engineering. 2014;**4**(5):312-324

[35] Rosandich TJ. Sports equipment and technology. The Sorts Journal. 2000;**20**

[36] Attom M. Information systems strategy and knowledge-based SMEs in the Australian biotechnology industry: Does IS need to reorient its thinking? In: 15th Australian Conference on Information Systems, Hobart. 2008

[37] LaMarco, N. (2019). Uses for computer in Business.Accessed online on 7/2/2020 from: https://smallbusiness.chron.com/uses-computers-business-56844.html.

[38] Pratt, M. (2013).Five essential hardware products for starting your business. Accessed online on 7/2/2020 from: https://www.business.org/it/hardware/essential-hardware-products-for-starting-your-business/

[39] Chandrasekaran K. Essentials of Cloud Computing. LLCCRC: Taylor & Francis Group; 2015

[40] Bester (2012). The business environment of sport organizations: A review. African Journal of Business Management, 6(46): 11530-11535. file:///C:/Users/ogwo/Downloads/The_business_environment_of_sport_organisations_A_.pdf

Chapter 11

Selecting a Laboratory Information System for Biobanks in Low- and Middle-Income Countries

Samuel Kyobe

Abstract

Biobanks in low-and middle-income countries need significant infrastructural support to meet ISBER Best Practices to support population-based genomics research. ISBER recommends a Biobank information management system that can manage workflows from biospecimen receipt to distribution. The H3Africa Initiative was set out to develop regional African Biobanks where Uganda, Nigeria, and South Africa were successfully awarded grants to develop state-of-the-art Biobanks. In this chapter, we review the African experiences, processes, and recommendations for information management systems for use in the low-and middle-income country context. We provide a balanced basis on which institutions can deliberate their decision between an out-of-the-box service and a commercial enterprise.

Keywords: LIMS, LMIC, Biobanks

1. Introduction

Biobanks require the linkage of high-quality material to data housed in a laboratory information management system (LIMS) which tracks each sample. A founding principle of H3Africa is to ensure that DNA (and possibly other clinical biological material) would be stored in Biobanks for future research purposes [1, 2]. The value of this material is partially determined by the associated phenotypic data. Management of this data and sample tracking in compliance with national and international best practices and ethical guidelines requires a refined data management system [3, 4]. According to the International Society for Biological and Environmental Repositories (ISBER) Best Practices, a computer-based inventory system should be in place to track the location and pertinent annotation of every specimen in the Biobank [3]. The system should also track significant events during a sample's existence from collection to destruction, including sample thaws, receipt and/or processing delays, processing, transfer of the sample within the repository, specimen distribution and return, and destruction [3–5]. These ISBER Best Practices, as well as others from around the world, are under consideration to codify an International Organization for

Standardization (ISO) norm as the basis for a new international accreditation program for Biobanks [6]. Several factors hinder the successful implementation of a Biobank laboratory information management system (LIMS) in low- and middle-income countries. This chapter discusses the experience of H3Africa Biobanks in the evaluation and setup of a sustainable Biobank information management system.

2. The importance of LIMS sustainability

Biobanks should develop strategies for long-term LIMS sustainability. Reliable and adequate sources of funding are key to the sustainability of LIMS. Commercial LIMS vendors require the payment of annual support fees, unlike open-source LIMS. The risk of liquidation needs to be considered as it impacts LIMS support and maintenance and the long-term viability of the Biobank. Biobanks should employ cost recovery measures for users to ensure sustainability.

3. LIMS harmonization in the case of H3Africa Biobanks

Before the H3Africa program, each Biobank had its LIMS that met their current needs. Since the commencement of the program, the Biobanks have either acquired new LIMS or upgraded their existing systems following a thorough LIMS assessment program. The Biobanks conducted a harmonization exercise to ensure the interoperability of the LIMS across the three sites. Pilot studies conducted between H3Africa Biobanks integrated data sharing and importation protocols through a pilot biospecimen and data exchange. Data exchange harmonization is essential if Biobanks are to operate efficiently in networks like H3Africa to support population genomics studies. Biobanks need to define an agreed sharable set of data and data formats for harmonization and interoperability to facilitate exchange. Pilot projects with virtual data transfer protocols were undertaken successfully suggesting that this harmonization has been effective.

LIMS general information
• What development tools have been used in the creation of the LIMS (e.g., Microsoft C#)?
• Is the system scalable and therefore suitable for small organisations as well as global corporations?
• What are the system requirements?
• Does the system allow concurrent operation of three or more database groups in the standard system?
System configuration/customisation
• Is your system configurable so that it can exactly meet our requirements? If yes what are the requirements? If no, what are the alternatives?
• Is the system configurable by the non-programmer and without use of special languages?
• Are user configuration changes supported by the vendor? Is there any extra cost?
User access/security
• What is the security access system in detail?
• Does your system password requirements comply with 21CFR Part 11?
Functionality—sample/work registration

• Does your system support the following types of registration: single sample; single sample with copy feature; batch registration with/without copy feature; bar-code support; registration templates; registration from external system and/or scheduler; spreadsheet style registration including data capture from Excel; can details of submitters and sample types be viewed from the registration screen; can reports, e.g., worksheets and labels be automatically generated
• Does system allow fields to be populated automatically by defining a default value within the system which can be overtyped by a user with suitable authority?
Functionality—sample receipt
• Is a sample receipt function provided with the system? Useful when some work is pre-registered prior to the availability of the samples. The receipt function is needed to track the arrival of the samples. Useful in checking whether all expected samples arrive in the laboratory and the time interval between registration and receipt for each sample can be measured. The sample turnaround time in the laboratory should be measured from the receipt date and time.
• Does sample receipt allow for single sample, multiple sample, batch sample and global sample receipt utilising bar-codes where needed?
Functionality—sample preparation
• Is a sample preparation function provided with the system? This would be used to indicate that samples must complete the preparation stage before they are ready for testing.
• Does sample preparation allow for single sample, multiple sample, batch sample and global sample preparation?
Functionality—result entry
• Does system include: result entry by sample—entry of any/all test results for a single sample; result entry by test—entry of one test result for multiple samples; result entry for multiple samples and multiple tests in spreadsheet style; result import from a variety of sources including files and instruments; viewing of results previously entered using same selection criteria as for selection of samples for result entry; viewing of test status.
Reporting
• What reporting tools does the system support?
• Is event triggered reporting—reports generated by sample status change, for example—included?
System support
• If I have a support question can I telephone a help-desk and immediately talk to a technical person familiar with my system? If yes, where would this person be located and what time can I call them? If no, what support scheme is in place?
• Does the vendor have global coverage if relevant? If so what are the support centre locations?
• Are new version upgrades supplied to customers at no cost?
• How much effort is typically needed for the implementation of an upgrade and what do I have to pay?
• Will an upgrade preserve my configuration or custom code as well as my data?
• Do you guarantee that we will always be able to upgrade to the next version?
Miscellaneous
• Does system allow storage of BLOB files, e.g., pictures, documents (consent forms) associated with a sample or a test?
• Can a document management capability be fully integrated within the system?
• Describe the sample tracking features of your LIMS. How is this used to monitor inventory for example?
• Can I have separate databases for different departments within my organisation (they may be in different locations)? Can these databases be configured differently?

Table 1.
Summarised checklist to consider while choosing a Biobank LIMS (a detailed checklist can be accessed via biorepository.h3africa.org).

3.1 Checklist for choosing an LIMS and H3Africa Biobank as model

H3Africa required that the three Biobank workflows would be harmonized through interoperable LIMS to enable data integration and exchange. Key elements that were considered while choosing a LIMS for the project included (a) customizability and usability; (b) interoperability with other LIMS; (c) access to revisions, updates, patches, and maintenance releases; (d) cost and access to technical support services; (e) maintenance and associated costs; (f) multiuser/multisite support; (g) robustness to handle large volumes of sample information; (g) security systems (audit trail, user roles, and privileges, etc.); and (h) type of (open-source or commercial) LIMS. Here we presented a summarized checklist (**Table 1**) we considered while choosing a Biobank LIMS; however, a detailed checklist can be accessed via Biobank.h3africa.org website.

4. IT infrastructure

In addition to high acquisition costs, commercial LIMS requires an IT infrastructure to fully support their function. IT infrastructure is a combined set of hardware (e.g., servers, computers), software (e.g., operating software), and network systems required to deploy and support the LIMS. As biospecimen numbers grow, there is a corresponding decrease in LIMS functionality such as very slow loading and processing speeds, which is linked to the supporting IT infrastructure [7]. Therefore, a LIMS infrastructure should have the ability to scale to meet the needs of the community it serves. Institutions that are unable to support a Biobank LIMS on their own should team with other institutions to develop and deploy a shared Biobank LIMS infrastructure. By using a shared infrastructure, each participating institution could maintain components of the infrastructure independently, while also collectively managing the entire Biobank architecture. This will not only provide an economic benefit but will also provide an environment for harmonizing complex, but still critical, components of a LIMS such as structured data files and data models, as well as standards for data transmission.

4.1 User support services

Commercial LIMS often require user support services such as customization, implementation assistance, annual licenses, maintenance, and update, although the need for these services may vary over time especially as users become experienced with the system. User support can be provided in several forms including telephonic support and on-site support. There is an additional cost to access such services, which may be minimized by conducting thorough initial training and license negotiation. Many commercial LIMS vendors have no support networks or offices in low- and middle-income countries. This increases the costs of user support services because of airfare, accommodation, and other attendant costs to access the services. In such circumstances, low- and middle-income country Biobanks should endeavor to use other remote access technology to access support such as public IP addresses that enable external access and manipulation of the LIMS. Unlike commercial LIMS, open-source LIMS systems do not have user support services, and the user must troubleshoot locally which can be challenging or impossible based on technical capacity.

5. Open-source LIMS and commercial LIMS options

Commercial LIMS are systems whose source code is developed for sale and requires authorization from vendors before licensed use. Open-source LIMS are systems whose source codes are made available for distribution at no cost [8]. During the H3Africa Biobank implementation phases, it was decided to implement commercial LIMS in support of the collections. Commercial LIMS are significantly more expensive upfront than open-source LIMS and are less flexible for end-user adaptations but do not require local expertise to support. Despite some features of modified open-source LIMS that might seem more applicable to the low- and middle-income country setting, there were significant concerns regarding the stability of such systems and the lack of standardization. Additionally, it was clear that the adaptation and maintenance of such a system would require highly specialized staff at each of the Biobanks and that this may create differences among the Biobanks which could cause potential delays in the interlaboratory transfer of data and material.

In general, both open-source and commercial LIMS have some benefits and drawbacks. Open-source LIMS are quite cheap and most are entirely free to acquire, easily customizable, and open to various platforms such as Linux, Window, or Unix. However, they are not secure to data hacking, fixing bugs requires IT expertise, and there are no user support services.

Commercial LIMS are highly secure and reliable and have dedicated user support available in the form of telephone and email support. Software updates are provided regularly most often free but sometimes at a user fee. Some are customizable and adhere to the ISO 20387:2018 [9]. However, they are expensive; in our experience, one commercial LIMS was purchased at US$120,000. Some are associated with annual renewal licenses or user fees which can be very prohibitive and affect the Biobank sustainability. One of the major drawbacks of commercial LIMS is discontinuation of the product without the consultation and consent of users leaving them in jeopardy.

6. Budgeting for LIMS in the H3Africa Biobanks

The functioning of Biobanks requires stability and continuity of the LIMS. In addition to the purchase of commercial licenses, some unforeseen expenses were encountered in the implementation of the commercial LIMS in H3Africa. These included training costs for staff and more complex hardware requirements, unlike for open-source LIMS. In the case of H3Africa, it was possible to budget for these contingencies, but these costs may become prohibitive for smaller Biobanks outside of a funded grant. In these cases, innovative solutions may include forming consortia with other facilities and purchasing a multiuser license. This may have the added advantage of enabling a shared forum for dealing with other problems. However, to keep each repository with separate views, specific role-based security would have to be implemented for each user's collections. It may also only be necessary to acquire specific modules within a commercial package to keep the costs lower.

Retaining and training staff to operate LIMS has presented some challenges for H3Africa. Some strategies were employed, including actively headhunting individuals with experience, ensuring that the LIMS manager feels integrated into the Biobank management structure, and training junior staff to ensure an adequate succession plan. Also, including training clauses within purchasing agreements have mitigated some of the risks of purchasing complex licenses, but ongoing training represents an essential need and should be part of a Biobank's quality management system. Discussion regarding formulating in-house training material for LIMS support is ongoing across all three H3Africa Biobanks to stay aligned with training and SOPs.

7. Conclusion

Developing a state-of-the-art Biobank requires considerable capacity and staff development including the acquisition of formal training, equipment, and software. Key among the Biobank infrastructure is a LIMS. Choosing a LIMS in low- and middle-income countries requires careful consideration of the various factors that could affect its successful and sustainable deployment and utilization. H3Africa Biobanks operating in a consortium have highlighted key factors and recommendations that affect successful LIMS implementation.

Author details

Samuel Kyobe
Department of Medical Microbiology, School of Biomedical Sciences, College of Health Sciences, Makerere University, Uganda

*Address all correspondence to: samuelkyobe@gmail.com

References

[1] H3Africa Consortium et al. Research capacity. Enabling the genomic revolution in Africa. Science. 2014;**344**(6190):1346-1348

[2] Human Heredity and Health in Africa (H3Africa). Harnessing genomic technologies toward improving health in Africa: Opportunities and challenges. In: H3Africa White Paper. National Institutes of Health; 2011. p. 49. Available from: https://h3africa.org/index.php/about/white-paper/

[3] Campbell LD et al. Development of the ISBER best practices for repositories: Collection, storage, retrieval and distribution of biological materials for research. Biopreservation and Biobanking. 2012;**10**(2):232-233

[4] Vaught J et al. The ISBER best practices: Insight from the editors of the third edition. Biopreservation and Biobanking. 2012;**10**(2):76-78

[5] Pitt K, Betsou F. The ISBER best practices self assessment tool (SAT): Lessons learned after three years of collecting responses. Biopreservation and Biobanking. 2012;**10**(6):548-549

[6] Betsou F, Luzergues A, Carter A, Geary P, Riegman P, Clark B, et al. Towards norms for accreditation of biobanks for human health and medical research: Compilation of existing guidelines into an ISO certification/accreditation norm-compatible format. Quality Assurance Journal. 2007;**11**:219-292

[7] Henricks WH. Laboratory information systems. Clinics in Laboratory Medicine. 2016;**36**(1):1-11

[8] Landgraf KM et al. Open-source LIMS in Vietnam: The path toward sustainability and host country ownership. International Journal of Medical Informatics. 2016;**93**:92-102

[9] International Organization for Standardization. Biotechnology—Biobanking—General requirements for biobanking (ISO Standard No. 20387:2018). 2018. Available from: https://www.iso.org/standard/67888.html